SPACECRAFT CHARGING BY MAGNETOSPHERIC PLASMAS

Edited by Alan Rosen
TRW, Inc.
Redondo Beach, California

Volume 47
PROGRESS IN
ASTRONAUTICS AND AERONAUTICS

Martin Summerfield, Series Editor-in-Chief
Princeton University, Princeton, New Jersey

Technical papers selected from the AIAA/AGU Symposium on Spacecraft Charging by Magnetospheric Plasmas, June 1975, subsequently revised for this volume.

Published by the American Institute of Aeronautics and Astronautics in cooperation with The MIT Press

American Institute of Aeronautics and Astronautics
New York, New York

The MIT Press
Cambridge, Massachusetts and London, England

Library of Congress Cataloging in Publication Data
Main entry under title:

AIAA/AGU Symposium on Spacecraft Charging by Magnetospheric Plasmas, Washington, D. C., 1975.
 Spacecraft charging by magnetospheric plasmas.

 (Progress in astronautics and aeronautics; v. 47)
 Includes bibliographies and index.
 1. Space vehicles——Electrostatic charging——Congresses. 2. Plasma
(Ionized gases)——Congresses. 3. Magnetosphere——Congresses.
I. Rosen, Alan. II. American Institute of Aeronautics and
Astronautics. III. American Geophysical Union. IV. Title. V. Series.
TL507.P75 vol. 47 [TL1492] 629.1'08s [629.47] 76-14821
ISBN 0-915928-11-6

Copyright © 1976 by
American Institute of Aeronautics and Astronautics

All rights reserved. No part of this book may be reproduced in any
form or by any means, electronic or mechanical, including photo-
copying, recording, or by any information storage and retrieval
system, without permission in writing from the publisher.

Table of Contents

Preface ... xi

I Spacecraft Charging Technology Programs ... 1

Spacecraft Charging Investigation: A Joint
Research and Technology Program ... 3
ROBERT R. LOVELL, N. JOHN STEVENS,
WAYNE SCHOBER, C. P. PIKE, AND
WILLIAM LEHN

Spacecraft Charging at High Altitudes: The
Scatha Satellite Program ... 15
D. A. MCPHERSON AND W. R. SCHOBER

A High-Time Resolution Spectrometer for 0.05
to 500 keV Electrons and Protons ... 31
B. SELLERS, F. A. HANSER, P. R. MOREL,
J. L. HUNERWADEL, A. L. PAVEL, L. KATZ,
AND P. L. ROTHWELL

II Environmental Interaction with Spacecraft ... 43

A Correlation Study Relating Spacecraft
Anomalies to Environmental Data ... 45
C. P. PIKE AND M. H. BUNN

Observations of Electrical Discharges Caused by
Differential Satellite-Charging ... 61
R. R. SHAW, J. E. NANEVICZ, AND R. C. ADAMO

Substorm-Induced Spacecraft-Charging Currents from
Field-Aligned and Omnidirectional Particles ... 77
J. L. VOGL, N. L. SANDERS, AND S. E. DEFOREST

Relationship between ATS-6 Spacecraft-Charging
Occurrences and Warm Plasma Encounters ... 89
D. L. REASONER, WALTER LENNARTSSON,
AND C. R. CHAPPELL

Spacecraft Potentials in a Substorm Environment ... 103
G. T. INOUYE

Simulation of the Plasma Sheath Surrounding a
Charged Spacecraft ... 121
P. L. ROTHWELL, A. G. RUBIN, A. L. PAVEL,
AND L. KATZ

Theory of the Spherically Symmetric Photoelectron
Sheath and Comparison with the ATS-6
Observation of a Potential Barrier ... 135
E. C. WHIPPLE JR.

Spacecraft Potential Control with
Electron Emitters ... 159
R. GRARD, A. GONFALONE, AND A. PEDERSEN

Active Control of Spacecraft Potentials
at Geosynchronous Orbit ... 169
R. GOLDSTEIN AND S. E. DEFOREST

III   Spacecraft Materials; Response to the Arcing Process — 183

Spacecraft Material Response to Geosynchronous Substorm Conditions — 185
D. K. HOFFMASTER AND J. M. SELLEN JR.

Surface Discharges on Spacecraft Dielectrics in a Scanning Electron Microscope — 213
K. G. BALMAIN, M. ORSZAG, AND P. KREMER

Spacecraft-Charging Studies of Voltage Breakdown Processes on Spacecraft Thermal Control Mirrors — 225
R. C. ADAMO AND J. E. NANEVICZ

Evidence for a New Discharge Mechanism for Dielectrics in a Plasma — 237
A. MEULENBERG JR.

Malter Discharges as a Possible Mechanism Responsible for Noise Pulses Observed on Synchronous-Orbit Satellites — 247
J. E. NANEVICZ AND R. C. ADAMO

Spacecraft-Charging Investigation for the CTS Project — 263
N. JOHN STEVENS, ROBERT R. LOVELL, AND VICTOR GORE

Spacecraft-Charging Analysis and Test for Environmentally Induced EMI — 277
A. KRAUSZ

Index to Contributors to Volume 47 — 289

Subject Index to Volume 47 — 290

Progress in Astronautics and Aeronautics

Martin Summerfield, Series Editor
PRINCETON UNIVERSITY

VOLUMES

1. Solid Propellant Rocket Research. 1960

   EDITORS
   Martin Summerfield
   PRINCETON UNIVERSITY

2. Liquid Rockets and Propellants. 1960

   Loren E. Bollinger
   THE OHIO STATE UNIVERSITY

   Martin Goldsmith
   THE RAND CORPORATION

   Alexis W. Lemmon Jr.
   BATTELLE MEMORIAL INSTITUTE

3. Energy Conversion for Space Power. 1961

   Nathan W. Snyder
   INSTITUTE FOR DEFENSE ANALYSES

4. Space Power Systems. 1961

   Nathan W. Snyder
   INSTITUTE FOR DEFENSE ANALYSES

5. Electrostatic Propulsion. 1961

   David B. Langmuir
   SPACE TECHNOLOGY LABORATORIES, INC.

   Ernst Stuhlinger
   NASA GEORGE C. MARSHALL SPACE FLIGHT CENTER

   J. M. Sellen Jr.
   SPACE TECHNOLOGY LABORATORIES

6. Detonation and Two-Phase Flow. 1962

   S. S. Penner
   CALIFORNIA INSTITUTE OF TECHNOLOGY

   F. A. Williams
   HARVARD UNIVERSITY

7. Hypersonic Flow Research. 1962

   Frederick R. Riddell
   AVCO CORPORATION

8. Guidance and Control. 1962

   Robert E. Roberson
   CONSULTANT

   James S. Farrior
   LOCKHEED MISSILES AND SPACE COMPANY

9. Electric Propulsion Development. 1963

   Ernst Stuhlinger
   NASA GEORGE C. MARSHALL SPACE FLIGHT CENTER

10. Technology of Lunar
    Exploration. 1963

Clifford I. Cummings and
Harold R. Lawrence
JET PROPULSION LABORATORY

11. Power Systems for Space
    Flight. 1963

Morris A. Zipkin and
Russell N. Edwards
GENERAL ELECTRIC COMPANY

12. Ionization in High-
    Temperature Gases. 1963

Kurt E. Shuler, Editor
NATIONAL BUREAU OF STANDARDS

John B. Fenn, Associate Editor
PRINCETON UNIVERSITY

13. Guidance and Control — II.
    1964

Robert C. Langford
GENERAL PRECISION INC.

Charles J. Mundo
INSTITUTE OF NAVAL STUDIES

14. Celestial Mechanics and
    Astrodynamics. 1964

Victor G. Szebehely
YALE UNIVERSITY OBSERVATORY

15. Heterogeneous Combustion.
    1964

Hans G. Wolfhard
INSTITUTE FOR DEFENSE ANALYSES

Irvin Glassman
PRINCETON UNIVERSITY

Leon Green Jr.
AIR FORCE SYSTEMS COMMAND

16. Space Power Systems
    Engineering. 1966

George C. Szego
INSTITUTE FOR DEFENSE ANALYSES

J. Edward Taylor
TRW INC.

17. Methods in Astrodynamics
    and Celestial Mechanics.
    1966

Raynor L. Duncombe
U.S. NAVAL OBSERVATORY

Victor G. Szebehely
YALE UNIVERSITY OBSERVATORY

18. Thermophysics and
    Temperature Control of
    Spacecraft and Entry
    Vehicles. 1966

Gerhard B. Heller
NASA GEORGE C. MARSHALL SPACE
FLIGHT CENTER

19. Communication Satellite
    Systems Technology. 1966

Richard B. Marsten
RADIO CORPORATION OF AMERICA

20. Thermophysics of Spacecraft and Planetary Bodies Radiation Properties of Solids and the Electromagnetic Radiation Environment in Space. 1967
    Gerhard B. Heller
    NASA GEORGE C. MARSHALL SPACE FLIGHT CENTER

21. Thermal Design Principles of Spacecraft and Entry Bodies. 1969
    Jerry T. Bevans
    TRW SYSTEMS

22. Stratospheric Circulation. 1969
    Willis L. Webb
    ATMOSPHERIC SCIENCES LABORATORY, WHITE SANDS, AND UNIVERSITY OF TEXAS AT EL PASO

23. Thermophysics: Applications to Thermal Design of Spacecraft. 1970
    Jerry T. Bevans
    TRW SYSTEMS

24. Heat Transfer and Spacecraft Thermal Control. 1971
    John W. Lucas
    JET PROPULSION LABORATORY

25. Communication Satellites for the 70's: Technology. 1971
    Nathaniel E. Feldman
    THE RAND CORPORATION

    Charles M. Kelly
    THE AEROSPACE CORPORATION

26. Communication Satellites for the 70's: Systems. 1971
    Nathaniel E. Feldman
    THE RAND CORPORATION

    Charles M. Kelly
    THE AEROSPACE CORPORATION

27. Thermospheric Circulation. 1972
    Willis L. Webb
    ATMOSPHERIC SCIENCES LABORATORY, WHITE SANDS, AND UNIVERSITY OF TEXAS AT EL PASO

28. Thermal Characteristics of the Moon. 1972
    John W. Lucas
    JET PROPULSION LABORATORY

29. Fundamentals of Spacecraft Thermal Design. 1972
    John W. Lucas
    JET PROPULSION LABORATORY

30. Solar Activity Observations and Predictions. 1972

Patrick S. McIntosh and
Murray Dryer
ENVIRONMENTAL RESEARCH LABORATORIES, NATIONAL OCEANIC AND ATMOSPHERIC ADMINISTRATION

31. Thermal Control and Radiation. 1973

Chang-Lin Tien
UNIVERSITY OF CALIFORNIA, BERKELEY

32. Communications Satellite Systems. 1974

P. L. Bargellini
COMSAT LABORATORIES

33. Communications Satellite Technology. 1974

P. L. Bargellini
COMSAT LABORATORIES

34. Instrumentation for Airbreathing Propulsion. 1974

Allen E. Fuhs
NAVAL POSTGRADUATE SCHOOL

Marshall Kingery
ARNOLD ENGINEERING DEVELOPMENT CENTER

35. Thermophysics and Spacecraft Thermal Control. 1974

Robert G. Hering
UNIVERSITY OF IOWA

36. Thermal Pollution Analysis. 1975

Joseph A. Schetz
VIRGINIA POLYTECHNIC INSTITUTE

37. Aeroacoustics: Jet and Combustion Noise; Duct Acoustics. 1975

Henry T. Nagamatsu, Editor
GENERAL ELECTRIC RESEARCH AND DEVELOPMENT CENTER

Jack V. O'Keefe, Associate Editor
THE BOEING COMPANY

Ira R. Schwartz, Associate Editor
NASA AMES RESEARCH CENTER

38. Aeroacoustics: Fan, STOL, and Boundary Layer Noise; Sonic Boom; Aeroacoustic Instrumentation. 1975

Henry T. Nagamatsu, Editor
GENERAL ELECTRIC RESEARCH AND DEVELOPMENT CENTER

Jack V. O'Keefe, Associate Editor
THE BOEING COMPANY

Ira R. Schwartz, Associate Editor
NASA AMES RESEARCH CENTER

39. Heat Transfer with Thermal
    Control Applications. 1975

    M. Michael Yovanovich
    UNIVERSITY OF WATERLOO

40. Aerodynamics of Base
    Combustion. 1976

    S. N. B. Murthy
    PURDUE UNIVERSITY

41. Communications Satellite
    Developments: Systems. 1976

    Gilbert E. LaVean
    DEFENSE COMMUNICATIONS
    ENGINEERING CENTER

    William G. Schmidt
    CML SATELLITE CORPORATION

42. Communications Satellite
    Developments: Technology. 1976

    William G. Schmidt
    CML SATELLITE CORPORATION

    Gilbert E. LaVean
    DEFENSE COMMUNICATIONS
    ENGINEERING CENTER

43. Aeroacoustics: Jet Noise,
    Combustion and Core Engine
    Noise. 1976

    Ira R. Schwartz
    NASA AMES RESEARCH CENTER

    Henry T. Nagamatsu
    GENERAL ELECTRIC RESEARCH AND
    DEVELOPMENT CENTER

    Warren C. Strahle
    GEORGIA INSTITUTE OF TECHNOLOGY

44. Aeroacoustics: Fan Noise and
    Control; Duct Acoustics:
    Rotor Noise. 1976

    Ira R. Schwartz
    NASA AMES RESEARCH CENTER

    Henry T. Nagamatsu
    GENERAL ELECTRIC RESEARCH AND
    DEVELOPMENT CENTER

    Warren C. Strahle
    GEORGIA INSTITUTE OF TECHNOLOGY

45. Aeroacoustics: STOL Noise;
    Airframe and Airfoil Noise. 1976

    Ira R. Schwartz
    NASA AMES RESEARCH CENTER

    Henry T. Nagamatsu
    GENERAL ELECTRIC RESEARCH AND
    DEVELOPMENT CENTER

    Warren C. Strahle
    GEORGIA INSTITUTE OF TECHNOLOGY

46. Aeroacoustics: Acoustic Wave
    Propagation; Aircraft Noise
    Prediction; Aeroacoustic
    Instrumentation. 1976

    Ira R. Schwartz
    NASA AMES RESEARCH CENTER

    Henry T. Nagamatsu
    GENERAL ELECTRIC RESEARCH AND
    DEVELOPMENT CENTER

    Warren C. Strahle
    GEORGIA INSTITUTE OF TECHNOLOGY

47. Spacecraft Charging by   Alan Rosen
    Magnetospheric Plasmas.   TRW, INC.
    1976

(Other volumes are planned.)

PREFACE

Spacecraft charging by magnetospheric plasma is a newly discovered space hazard that can virtually destroy spacecraft in Earth orbit if they are not properly designed. At least one catastrophic spacecraft failure and innumerable malfunctions and system interruptions can be attributed to this phenomenon. A large fraction of the available information has been presented in symposia and in scattered articles in various technical, scientific, and engineering journals. The symposium on "Spacecraft Charging by Magnetospheric Plasmas," held in Washington, D.C. on June 16, 1975 and cosponsored by the AIAA and the American Geophysical Union, was an attempt to consolidate all that is known on the subject. This volume is a collection of the manuscripts presented at the symposium, subsequently revised and brought up to date for presentation herein. It is divided into three chapters so designed that each chapter may be read separately as a self-contained unit without reference to the other chapters.

Spacecraft charging up to the multikilovolt range was discovered in 1971 and has since received the attention of a large number of scientists and engineers; many of them are represented as authors in this volume. Their work may be summarized by the conclusion that the following sequence of events accounts for the spacecraft failures and anomalies observed in orbit:

1) Immersion of a spacecraft in a substorm plasma.

2) Differential charging of component parts of the spacecraft to a high voltage difference.

3) Generation of a vacuum arc when the voltage stress level exceeds the breakdown potential of the material.

4) Irradiation of spacecraft components by the electromagnetic interference (EMI) waves associated with the vacuum arc.

5) Induction into onboard electronic circuitry of a transient pulse of sufficient magnitude to activate the circuit or burn out some of its components.

There are broad implications resulting from these phenomena, not only with respect to the direct engineering problem of spacecraft design but also with respect to our scientific knowledge of the ionized mediums in planetary and interplanetary space, and these are discussed in the three chapters of this volume.

Chapter 1, Spacecraft Charging Technology Programs, describes the research programs instituted by the National Aeronautics and Space Administration and the U.S. Air Force in response to this problem.

Chapter 2, Environmental Interaction with Spacecraft, describes the effects of the space environment on spacecraft. The chapter includes the observed malfunctions of spacecraft in orbit and their correlation with environmental substorm events, a description of that portion of the space environment that is responsible for charging spacecraft to the multikilovolt level, an analysis of the interaction that produces multikilovolt potentials, and some passive and active control methods for modifying these potentials.

Chapter 3, Spacecraft Materials; Response to the Arcing Process, describes the relevant material properties and the mechanisms for the generation of arcs on spacecraft and, in addition, methods and procedures for testing spacecraft for their susceptibility to the associated arcing phenomena.

Each chapter of this volume is introduced with a summary of its contents and a description of the major problems in that area.

The problem of spacecraft charging has not yet been solved—we look to solutions in the following areas:

1) A standard environmental specification to cover the geomagnetic substorm environment needs to be generated for spacecraft designers and fabricators.

2) General design requirements and criteria need to be standardized.

3) A logical nondestructive spacecraft test program needs to be developed to assure adherence to the design standards.

4) The housekeeping data system of every operational spacecraft at synchronous altitude should carry simple monitors to determine substorm induced noise (RFI) and potential differences.

5) A laboratory test program should be undertaken in the selection of spacecraft materials.

6) An inflight test program should be undertaken for evaluation and testing.

7) Additional environmental data should be obtained.

We hope that this volume is useful in directing future research on the physics of the planetary and interplanetary media as well as on spacecraft design, and that it will help point development efforts toward the necessary solution.

Alan Rosen
TRW Defense and Space Systems

January 1976

Chapter 1—Spacecraft Charging Technology Programs

A number of organizations, concerned with the problem of environmentally induced anomalies on spacecraft, have instituted programs to deal with various aspects of the problem. The program encompassing the largest scope is the joint Air Force-NASA program that has been established to investigate the spacecraft charging phenomenon. The first paper in this chapter describes this program and outlines the laboratory and flight test research projects that will be undertaken within the next four years. The space measurement program entitled Spacecraft Charging at High Altitudes (SCATHA) and being performed by the Air Force is described in the next paper. One of the instruments to be flown on the SCATHA satellite is described in the third paper as an example of the type of instrumentation that is required to solve the problems associated with spacecraft charging by environmental plasma.

There are, in addition to the programs described by the papers in this chapter, a significant number of smaller programs sponsored by a variety of organizations, all aimed at solving various aspects of the spacecraft charging problem. TRW Defense and Space Systems has had a number of programs aimed at studying the environmental interaction with spacecraft and the spacecraft arcing process. Special emphasis was given to studies of procedures for spacecraft testing and methods of designing spacecraft so as to minimize the effects of the charging environment. These are described in Chapters 2 and 3. The Air Force Cambridge Research Laboratory has undertaken experimental and theoretical environmental studies and correlation studies to establish that the space environment was in fact the cause of many of the observed anomalies. A number of programs to study the behavior of materials under conditions of charge buildup are now underway. Some of the most noteworthy are at the Air Force Systems Command, Aeronautical Systems Division, Wright-Patterson Air Force Base; the University of Toronto; Lincoln Laboratories; and Stanford Research Institute. Programs to study new discharge meahanisms have been undertaken by Comsat Laboratories and Stanford Research Institute. The papers presented in Chapters 2 and 3 cover all these areas in some detail and generally outline anticipated future developments.

SPACECRAFT CHARGING INVESTIGATION:
A JOINT RESEARCH AND TECHNOLOGY PROGRAM

Robert R. Lovell[*] and N. John Stevens[†]
NASA Lewis Research Center, Cleveland, Ohio

Wayne Schober[‡]
Space and Missile Organization, El Segundo, Calif.

C. P. Pike[§]
Air Force Geophysics Laboratory
Hanscom Air Force Base, Mass.

and

William Lehn[¶]
Air Force Materials Laboratory
Wright-Patterson Air Force Base, Ohio

Abstract

A jointly planned U.S. Air Force-NASA program has been established to investigate the spacecraft charging phenomenon that has caused electronic anomalies in satellites in geosynchronous orbits. The objectives of this program are to provide design criteria, techniques, and test methods to insure control of absolute and differential charging of spacecraft surfaces. These objectives will be updated continuously over the next four years as data become available from the combined contractual and in-house programs. The geosynchronous altitude environment will be defined, ground and flight tests will be conducted, and materials and charge control techniques will be developed as required. The ultimate output of the program will be a space-

---

Presented as paper SA44, at the 1975 Spring Annual Meeting American Geophysical Union, Washington, D.C., June 18-19, 1975.
*Chief, Systems Engineering Branch.
†Aerospace Engineer.
‡Experiment Manager, Scatha Program.
§Physicist (Physics of the Atmosphere).
¶Technical Manager for Coatings.

craft charging design criteria and test specification document. The program will be coordinated by a spacecraft charging program review group which has both Air Force and NASA representation.

Introduction

During the early 1970's, several synchronous satellite programs began to report spurious switching activity near the local midnight region of the satellite orbits. These satellites were designed and developed in the late 1960's prior to any knowledge of an adverse space environment at synchronous altitudes. ATS-5 data[1,2] revealed the existence of transient particle fluxes of higher-than-expected energies in the local evening and midnight sectors of the orbit. Later work[3] indicated that the experimental data from ATS-5 spectrometer could be explained only by the assumption that the satellite was being charged by these particle fluxes. When the satellite was in Earth's shadow, it was found that the satellite ground was depressed to values as large as - 10 kV. A model of the environment was developed which explained the transient flux as the result of solar storm particles injected into Earth's magnetic field. This is referred to as a geomagnetic substorm. Correlation between the existence of a substorm and the spurious electronic behavior on several satellites has been obtained.[4] Simple models of the mechanism for spacecraft charging and discharging in a substorm environment have been postulated, and some engineering fixes have been incorporated in recently launched operational satellites, but the anomalies still persist.

A satellite immersed in an ambient plasma will come into electrical equilibrium with that plasma by developing surface charges of the proper sign and magnitude to reduce the net current between the satellite and the ambient plasma to zero. The net current consists of currents from the environmental flux, secondary and backscattered electrons and ions, and by photoelectrons from any illuminated areas. In a neutral low-energy plasma, as would have been expected at synchronous altitude, the surface should remain close to the space plasma potential. In the transient fluxes measured by ATS-5 and 6, the surfaces can charge to 10-20 keV negative when in the dark and to a few hundred volts negative when in the sun.

It should be pointed out that the resulting electric fields generated by the charged surface are complicated and may affect the basic measurements leading to this model. When a satellite is completely in the Earth's shadow and charged uniformly, then some kind of active or passive emitter will be required to reduce the potential to near zero in order that ambient flux density and energy can be measured.

A satellite with parts in the sun and parts in the shade can be expected to be charged differentially. The resistance and capacitance between the various parts, as well as the dynamic characteristics of the ambient flux and spacecraft spin rate, will determine the charging/discharging rates. If two adjacent parts of a satellite are charged to a multikilovolt differential, then the stress may be great enough to cause breakdown of the dielectric between the parts or arcing. The resulting discharge will emit electromagnetic pulses with energies proportional to the capacitance and voltage of the discharging circuit. The advanced operational satellites now being placed into geosynchronous orbit make extensive use of low-level logic in command circuits. This logic is susceptible to disruption by the arc-generated electromagnetic interference. Hence, the discharges are the most likely cause of the anomalous behavior just described.

In addition to causing the switching anomalies, the discharges also are known to cause degradation of the thermal control properties of the surfaces. This results in the rise in satellite component temperatures. Since the lifetime requirements of geosynchronous satellites are increasing from the present 2 to 10 yr in the near future, it is necessary to prevent this degradation to have a stable thermal control system for this increased lifetime.

As synchronous satellites become more numerous and complicated and as the solar activity increases over the next few years, it can be expected that the spacecraft charging anomalies will become more frequent and more serious. Because of this, it was decided that the resources of the U.S. Air Force and NASA should be combined to conduct a joint program. The description, the approach, and the goals of the program are given in the following paragraphs.

Spacecraft Charging Investigation

Objectives

The overall objective of this investigation is to provide the design criteria, materials, techniques and test methods to insure control of absolute and differential charging of space-

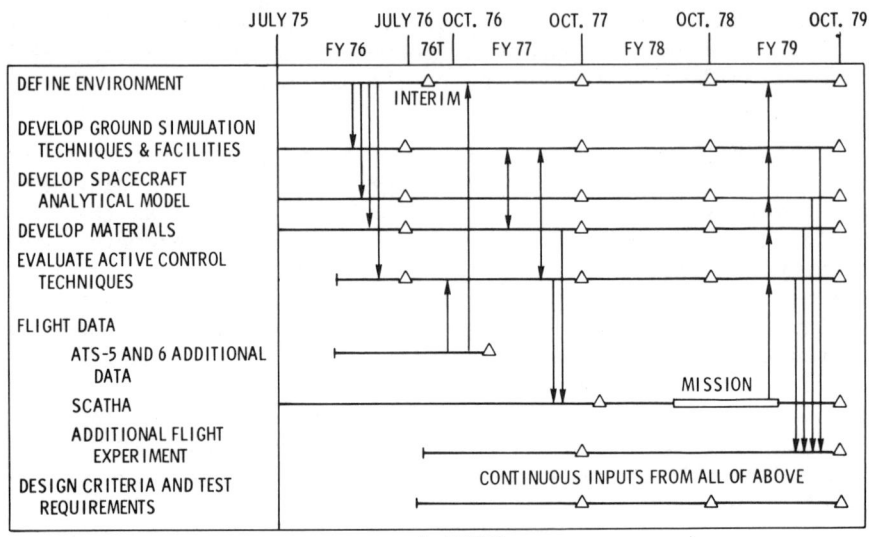

Fig. 1 Spacecraft charging investigation.

craft surfaces. This objective will be accomplished in the following series of steps:

1) The environment will be defined and a model established.
2) Ground facilities will be developed to simulate this environment and test techniques established to study the charging phenomena.
3) A spacecraft analytical model will be developed so that the interactions between the environment and the spacecraft surfaces can be evaluated and predicted.
4) Spacecraft materials will be modified or new materials will be developed to control the charging of the surfaces.
5) Active control techniques will be evaluated as a means to minimize surface charging.
6) Flight experiments will be conducted.
7) Design criteria and test specifications will be developed from the information thus gained.

The spacecraft charging investigation currently is planned as a 4-yr program, as shown in Fig. 1. There will be considerable interaction among the various elements of this program. The interconnecting lines on this figure are to show this interaction and are not an indication of the time that the exchange is to take place.

The present model of the environment will be used in the development of the ground facilities, in the establishment of the spacecraft analytical model, and as a guide in the development of new or modified materials. The environmental model will be improved by reducing the available data from space flight programs, correlating it to ground-based environmental observations, and obtaining additional data from experiments on ATS-5 and 6. As the environmental model is improved, the information will be supplied for upgrading the ground facilities, improving the spacecraft analytical model, and aiding in the materials development.

The SCATHA (Spacecraft Charging at the High Altitudes) satellite[5] is being built by the Air Force under the Space Test Program. The payload for this satellite will be composed of space experiments provided by the Air Force, NASA and the Navy. This satellite will be used in the spacecraft charging investigation to obtain flight data on the environment and to obtain the materials response to that environment. This information will be incorporated into the appropriate parts of the charging investigation. Additional flight experiments, if necessary, will be developed as the program progresses and will take advantage of the techniques, information, and materials that will come from this investigation. The flight experiments can be as simple as harness noise detectors and differential voltage sensor or as complex as several dedicated spacecraft.

As stated previously, the ultimate output of the joint AF-NASA program is the design criteria and test requirement documents for control of absolute and differential charging of spacecraft surfaces. These documents will utilize all the information obtained in the elements of this investigation. The documents will be issued in a preliminary form early in the program and upgraded periodically throughout the program.

Approach

The approach to be used to accomplish the objectives of this program was devised by representatives of NASA, SAMSO, the A.F. Geophysics Laboratory and the A.F. Materials Laboratory. The Lewis Research Center was established as the NASA lead center and the A.F. lead was SAMSO. The representatives of the various agencies were named as the members of the Program Review Group.

In the following paragraphs the technical approach to be used in the program will be discussed. For each element of the investigation, the approach will be given, the interactions with the other elements will be specified and the lead agency conducting the work will be identified.

Define Environment. There are two tasks to be accomplished under this element of the investigation:

Task 1) Climatological atlas of the synchronous environment. The available space and ground observatory data will be cataloged to define the space environmental conditions that exist at the geosynchronous altitudes for both the geomagnetic substorm and quiet conditions. The prime source of data for this work will be the existing results from the ATS-5 and 6 experiments. This information will be correlated with the data from ground observatories around the world to give a comprehensive picture of the environment. It is planned that an atlas will be published as an interim report in September 1976 and issued as a report in September 1977. The atlas will be updated by addendums as data are obtained from later flight experiments (such as SCATHA). The information obtained for this atlas will be used to support the other program elements. Therefore, there will be interactions between this work and the development of ground facilities, the establishment of the spacecraft analytical model, the development of materials and devices, and the design of flight experiments. The work to develop the atlas will be directed by the U.S. Air Force Geophysics Laboratory.

Task 2) Conduct additional ATS-5 and 6 environmental studies. Additional space environment data will be obtained from the ATS-5 and 6 particle and fields experiments to determine the present environmental conditions for quiet times and for substorms. The main device that will be utilized will be the University of California at San Diego (UCSD) auroral particles experiment. The data obtained in this task will be incorporated into the Climatological Atlas (task 1). As part of this task, a study of active control devices in space will be undertaken. This study will use the neutralizers of the cesium electric thrusters on the ATS-5 and 6 satellites. The objectives of this study are to determine the ability of the electron sources to control the spacecraft ground potential and to investigate the voltage sheath surrounding the spacecraft. This work will be directed by the NASA Lewis Research Center.

Develop Ground Simulation Techniques and Facilities. The task to be accomplished under this element of the investigation is described below. This work will be directed by the NASA Lewis Research Center:

Task 1) Develop ground simulation facilities. The experimental determination of the response of the various spacecraft materials is dependent upon a reasonable simulation of the environment. The existing facilities will be reviewed and their performance compared with the present knowledge of the environment. The environmental simulation will be improved as additional data become available. In addition, as better information on the spacecraft charging environment is obtained, the facilities will be upgraded to match the environmental conditions. The instrumentation required to monitor the environment in the ground facility as well as to study the rapid transients involved in the discharges will be developed. Facilities also will be developed to study the effect of the substorm environment on scale models of spacecraft. This information will be correlated to the performance of satellites in orbit. As part of this task, devices and instruments will be developed to test spacecraft in air to determine the susceptibility of the spacecraft systems to discharge induced anomalous behavior.

Development of a Theoretical Model of the Charging of Spacecraft. There are three major tasks that will be accomplished under this element of the investigation. These tasks are described below:

Task 1) Development of a spacecraft theoretical model. A spacecraft theoretical model will be developed so that studies of the interactions between the spacecraft surfaces and the environment can be conducted. This model will be an extension of the existing simplified models. The goal is to have a realistic, three-dimensional model of a generalized spacecraft. All applicable information obtained in environmental and materials studies will be incorporated into this model. This work will be directed by the U.S. Air Force Geophysics Laboratory.

Task 2) Materials characterization. There are two parts to this task: the engineering property evaluation and the determination of the classical physical properties:

1) Under the engineering property evaluation the behavior of the various spacecraft surfaces exposed to the simulated space substorm environment will be determined. The properties of interest in this study are those related to discharges on or through the test material surfaces. Therefore, the leakage current through and along the surface, the threshold voltages for discharges, and the discharge characteristics must be measured for the various spacecraft surfaces. These parameters must be known as a function of incident particle energy, flux, and species, as a function of material thickness and geometry, and as a function of sample temperature and illumination. This engineering property evaluation will be directed by the NASA Lewis Research Center. The data obtained in this evaluation will be incorporated into the spacecraft analytical model.

2) Under the determination of the classical properties, the basic mechanisms of energetic electron interactions with the various spacecraft materials will be determined. Theoretical and experimental studies will be performed to determine electron induced conductivity, secondary electron emission, charge storage, and injected charge distribution. The study will begin with a literature search to determine the information that is available in these areas and then develop into a laboratory program to obtain the data that is not available. This work will be directed by the A.F. Geophysics Laboratory.

Task 3) Study of charging and discharging mechanisms. The fundamental charge and discharge mechanisms involved in the interaction between spacecraft surfaces and the geomagnetic substorm environment will be studied under this task. Analytical models of these mechanisms will be developed. Particular attention will be devoted to defining the basis relationship between the charging and discharging mechanisms and the basic material properties, surrounding magnetic and electric fields, temperature, and sample geometry. This task will be both analytical and experimental, and will be directed by the NASA Lewis Research Center. The computer modeling of these mechanisms will be compatible with the computer model of the spacecraft developed by the U.S. Air Force Geophysics Laboratory so that both can be included in the spacecraft analytical model.

Develop New or Modified Materials. The task to be accomplished under this element of the investigation is described below. This task will be directed by the U.S. Air Force Materials Laboratory.

Task 1) Development of conductive spacecraft materials. The control of absolute and differential charging of spacecraft cannot be affected without the development of new and improved or modified materials which will provide electrical continuity over the surface of the spacecraft. The materials photoemission, secondary emission, thermooptical, physical and electrical properties in the space vacuum environment both in the presence and absence of electrical stress and ultraviolet, electron, and particulate radiation are important to the achievement of charge control. The materials must be stable or have predictable response to exposure to the space environment for long periods of time. The development of these new or modified materials will be undertaken. The materials of interest include conductive polymers, paints, transparent films and coatings as well as fabric coating interweaves. The engineering properties of these materials, as defined in the materials characterization section above, will be evaluated in the facilities developed by the NASA Lewis Research Center. Electrical,

SPACECRAFT CHARGING INVESTIGATION

thermal, and optical properties, as well as the basic physical structure of the materials will be measured by the Air Force Geophysics Laboratory.

Evaluate Active Control Techniques. The task to be accomplished under this element of the investigation is described below. This task will be directed by the NASA Lewis Research Center:

Task 1) Active charge control methods. An investigation into active means to control spacecraft charging will be undertaken. These methods will include utilization of various electron, plasma, ion, and photon emitters, as well as geometric control techniques such as grids and pinholes. The investigation will determine the feasibility of use of such devices, the difficulties in operating the devices, the impact on satellite performance, and the probability of successfully limiting or controlling the surface charge. A study of the ability of active control devices to control the spacecraft potential will be conducted in the space environment using the electron emitters on the ATS-5 and 6 satellites. This study has been described under the subsection "Define Environment," task 2.

Conduct Flight Experiments. Under this element of the spacecraft charging investigation, there are three tasks to be accomplished:

Task 1) Obtain additional ATS-5 and 6 data. The purpose of this work is to obtain data that can be used to define the space environment. This work already has been described in the define environment element of this investigation in task 2.

Task 2) Evaluate the SCATHA information. The SCATHA program is an ongoing Air Force project to build and launch, in mid 1978, a satellite to study the spacecraft charging phenomena and at the same time measure the substorm-generated environmental parameters. This satellite program is not an integral part of the spacecraft charging investigation, but the information obtained in this space flight will be very useful to verify and upgrade the results of the ground-based investigation. The engineering experiments on SCATHA will measure the materials response to the environmental charging and the electromagnetic interference from a discharge. These data can be used to substantiate the design criteria and test specifications that will be generated as the primary objective of the charging investigation. The SCATHA satellite will carry numerous instruments to measure the environmental parameters over a broad energy range. This information will be used to upgrade the environmental model developed from the ATS-5 and 6 data.

The spacecraft charging analytical model developed in the charging investigation can be used to predict the performance of the SCATHA satellite. By comparing the predictions to the actual flight data, the analytical model can be calibrated for future use. The SCATHA satellite is well suited for this role since it is designed and instrumented to study the spacecraft charging phenomena. The A.F. Materials Laboratory Thermal Control Coatings/Contamination, Synchronous Orbit Experiment on SCATHA will measure spacecraft contamination and the effects of spacecraft surface potential on spacecraft contamination as well as the performance of selected state of the art and newly developed conductive spacecraft thermal control materials in the synchronous orbit environment. The results of this and the engineering experiments will be used to guide the selection and development of improved materials and to relate and correlate spacecraft data to ground test results. There will be active and passive charge control systems on the SCATHA satellites. The data from the operation of these systems will verify the concepts and test data obtained in the ground program.

Task 3) Conduct additional flight experiments. This topic has been placed in the spacecraft charging investigation in the realization that as this program develops, additional flight data will be required. The purpose of this work will be to recommend the additional flight experiments that are required. At this time the possible flight experiments that may be considered range from simple environmental effects experiments (arc counters and differential voltage detectors) that should be carried on all synchronous orbiting satellites to another full satellite.

Prepare Design Criteria and Test Specifications. The work to be done under this element of the spacecraft charging investigation involves the documentation that will be published to satisfy the primary objectives of this program. These documents include the Climatological Atlas, which will define the environment and the spacecraft analytical model. There are two additional tasks to be accomplished under this element of the investigation and these tasks are described below:

Task 1) Preparation of a spacecraft design criteria document. This document will utilize the information gained in this investigation to set forth spacecraft design guidelines that should be used to avoid or control the effects of environmental charging. The document will contain material characteristics and discharge characteristics. The document will be issued as an interim report late in 1976 and upgraded periodically throughout the charging investigation. A final report will be issued at the end of the investigation. This work will be directed by the NASA Lewis Research Center.

Task 2) Preparation of a test specification. This document will utilize the information gained in this investigation to set forth spacecraft ground test requirements that should be used to insure that the spacecraft systems are immune to environmentally induced discharges. This specification will cover component, subsystem, and spacecraft level testing. The specification will be issued as an interim document in late 1976 and upgraded as the investigation progresses. A final specification will be issued at the conclusion of the charging investigation. This work will be directed by the Space and Missile Systems Organization.

Program Direction

This program will be directed by a joint U.S. Air Force - NASA spacecraft charging program review group. The function of this group is to coordinate the work being done by industry, universities, NASA, and the Air Force, and to review the progress of the program periodically. This group will recommend changes as required to satisfy the objectives of the program. This group will issue biannual reports to the NASA and A.F. Space Technology Group. The membership in this review group includes James Lazar, NASA Headquarters; Robert Lovell, NASA Lewis Research Center; Wayne Schober, SAMSO; Paul Metzger, Aerospace; Charles Pike, U.S. Air Force Geophysics Laboratory; and William Lehn, Air Force Materials Laboratory. The Air Force point of contact is Wayne Schober, and the NASA contact is Robert Lovell.

Concluding Remarks

A jointly planned U.S. Air Force - NASA program has been established to conduct a research and technology investigation into the spacecraft charging phenomenon. The objective of this program is to provide design criteria, materials, techniques, and test methods to insure control of absolute and differential charging of spacecraft surfaces. The program will consist of a combined contractual and in-house efforts aimed at understanding the spacecraft charging phenomena and relating spaceflight data to ground test results. There will be analytical programs to define the space environment and to model the spacecraft interaction with this environment. There will be experimental programs to develop ground facilities to simulate this environment, to determine the response of spacecraft materials to this environment, and to develop new or modified materials.

There will be flight programs to obtain space environment data and to evaluate the response to the substorm environment. The ultimate output of the program will be a spacecraft charging design criteria and test specification documents. The program will be coordinated by a spacecraft charging program review group, which will incorporate into this investigation the requirements of both NASA and the U.S. Air Force.

References

[1] Sharp, R. D., Shelley, E. G., Johnson, R. G., and Paschmann, G., "Preliminary Results of a Low Energy Particle Survey At Synchronous Altitude," Journal of Geophysical Research, Vol. 75, 1970, p. 6092.

[2] DeForest, S. E. and McIlwain, C. E., "Plasma Clouds in the Magnetosphere," Journal of Geophysical Research, Vol. 76, June 1971, pp. 3587-3611.

[3] DeForest, S. E., "Spacecraft Charging at Synchronous Orbits," Journal of Geophysical Research, Vol. 77, Feb. 1972, pp. 651-659.

[4] Fredricks, R. W. and Scarf, F. L., "Observations of Spacecraft Charging Effects in Energetic Plasma Regions," Photon and Particle Interactions with Surfaces in Space, edited by R. J. L. Grard, D. Reidel Publishing Co., Dordrecht-Holland, 1973, pp. 277-308.

[5] McPherson, D. A., Cauffman, D. P., and Schober, W., "Spacecraft Charging at the High Altitudes - The SCATHA Satellite Program," AIAA Paper 75-92, 1975, Pasadena, Calif., published elsewhere in this volume.

SPACECRAFT CHARGING AT HIGH ALTITUDES: THE SCATHA SATELLITE PROGRAM

D. A. McPherson[*]
The Aerospace Corporation, El Segundo, Calif.

and

W. R. Schober[†]
Space and Missile Systems Organization, El Segundo, Calif.

Abstract

Satellites at synchronous altitude exhibit unexplained behavior in the operation of electronic circuits and in the performance of thermal controls. A possible explanation for this behavior is that satellites can be charged to large negative voltages by energetic electrons in the space environment. A space measurements program entitled Spacecraft Charging at High Altitudes (SCATHA) has been formulated to determine the characteristics of the charging process, to measure the response of the satellite when charging occurs, and to evaluate the utility of various corrective techniques which can minimize differential charging on the satellite. These data will be used to formulate spacecraft charging specifications, test procedures, and models.

I. Introduction

Satellites in synchronous orbit are observed to exhibit unexplained operation of electronic circuits and degradation

---

Presented as Paper 75-92 at the AIAA 13th Aerospace Sciences Meeting, Pasadena, Calif., Jan. 20-22, 1975.
[*]Presently with Science Applications, Inc., El Segundo, California.
[†]Presently with Aerojet ElectroSystems Company, Azusa, California.

of thermal control properties. A possible cause for some of this behavior may be that satellites can be electrically charged by electrons in the space environment. The absolute potential between the satellite and the surrounding space plasma can exceed thousands of volts. The potential distribution about the satellite is a function of photoelectric emission from the different surface materials, of plasma flux to the satellite, and of the satellite geometry. The plasma flux is dependent upon the number density and energy distribution of plasma particles, particularly the electrons, because of the high electron mobility relative to the ions.

A satellite measurements program, SCATHA, has been formulated to measure the characteristics of the charging process, to determine the response of the satellite to charging, and to evaluate techniques which may correct the problem. The SCATHA satellite is scheduled for launch in mid-1978.

II. Background

Table 1 provides a summary of certain unexplained events observed on various Department of Defense and commercial satellites. A thorough study of all synchronous satellites for unexplained behavior has not been made. However, for each satellite investigated, some sort of unexplained behavior was found.

There are many examples that demonstrate the dependence of satellite unexplained behavior on geophysical parameters such as local time and geomagnetic activity. Figure 1 shows the dependence on local time of logic circuit upsets observed in several satellites. Figure 1 is a view of the equatorial plane divided into local time segments. Noon is at the top of Fig. 1 and midnight is at the bottom. A synchronous-altitude satellite will, of course, travel through all local times once each day. The data points show the local time at which spurious changes in logic circuits occurred for several different satellites. The radial position of each point is irrelevant. The exact times of the DSCS II RGA upsets are not known. The times during which they could have occurred are indicated by the connected squares.

It is clear from Fig. 1 that the events occur predominantly between midnight and dawn in satellite local time. Excluding the DSCS II RGA upsets, the probability of observing 31 of the other 47 events in the midnight-to-dawn quadrant is 3.3×10^{-9}, if there were not correlation with local time.

In summary, five different satellite systems involving 19 different flights have been investigated. All vehicles

Table 1. Synchronous satellites exhibit unexplained performance

Satellite	Action
Defense Satellite Communication System II satellites 1, 2, 3, 4	Reset generator assembly trigger (satellites 1 and 2 only) Power system failure (satellite 1 only) Gain control logic switching
Defense Support Program flights 1, 2, 3, 4	Thermal control degradation Sensor data noise Control circuits switching
Intelsat III five satellites	Mechanical despun assembly switching
Intelsat IV flights 3, 4, 5, 7	Erroneous operation of attitude control system
Telesat flights 1, 2	Telemetry logic switching

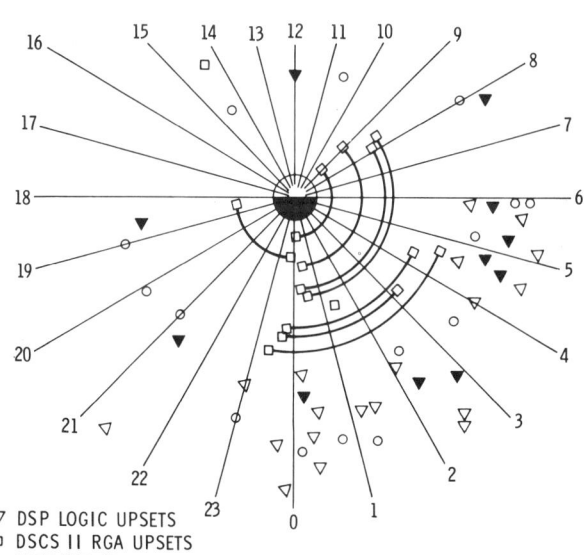

Fig. 1 Local time dependence of circuit upset for several DoD and commercial satellites.

have exhibited some type of behavior which cannot be explained and which generally depend upon geophysical parameters.

III. Spacecraft-Charging

DeForest has determined that the surface of the ATS-5 satellite could be negatively charged to hundreds of volts when the satellite was in sunlight and thousands of volts when the satellite was eclipsed.[1] Figure 2 illustrates the charging phenomenon. There are three processes contributing fluxes of charged particles to a satellite. Photoelectrons are emitted from surfaces illuminated by the sun. The current density emitted is around 5×10^{-5} A/m² and depends upon the surface material. The second contribution is the flux of electrons and ions from the space plasma to the satellite. The magnitudes of these fluxes are dependent upon the charged particle density and energy distribution. The third contributor is secondary emission of electrons resulting from incident energetic electrons. As a result of these three processes contributing to the charged particle fluxes, a potential distribution exists about the satellite so that the net current to the satellite is zero. Because the various particle fluxes are anisotropic and because the potential distribution is dependent upon satellite geometry, the potential distribution about the satellite can be highly asymmetric.[2]

At altitudes beyond 3 Earth radii, the ambient plasma density is generally less than 10^6 m⁻³.[3] In this case, as indicated in Fig. 2, the photoelectric current density is much greater than the flux of electrons from the plasma, and the total satellite potential is dominated by photoelectric emission. For example, it has been determined that the Vela satellites, which are at 111,000 km, can be charged positively by photoelectric emission. Potentials up to 100 v relative to the space plasma have been observed on the Vela satellites.[4]

At altitudes less than 2 Earth radii, or 13,000 km, the plasma density generally is greater than 10^9 m⁻³, and the temperature is similar to the ionospheric source of this plasma, namely, less than 1 ev. In this case, the flux of plasma electrons to the satellite is much greater than the photoelectron flux, and the satellite potential is slightly negative to the space plasma.

In summary, as one proceeds from low to high altitudes, the satellite potential will vary from negative to positive as the plasma electron flux becomes less then the photoelectric flux. An analysis of data from electric field probes

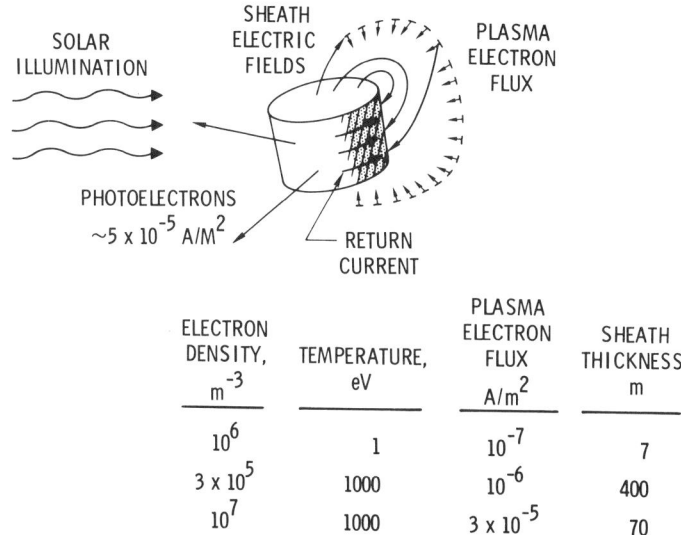

Fig. 2 Model of satellite differential charging by photoelectric emission and plasma flux.

on the Explorer 45 satellite, which had a 22-km-alt perigee and a 26,900-km-alt apogee, shows clearly the shift from a plasma dominated potential distribution to a photoelectric dominated potential distribution with increasing altitude.[2]

At synchronous altitude, the spacecraft potential will be dominated generally by photoelectric emission. It appears that, at times, the electron density and energy spectrum of the surrounding plasma are sufficient to charge the shadowed insulated parts of the satellite to hundreds and perhaps thousands of volts. If the surfaces of the satellite are insulated, which is usually the case, large differential voltages can be developed between the exposed shadowed surface and the satellite frame beneath the surface. The electric fields across dielectrics can be large enough to break down the material or to generate vacuum arcs at the edge of the material.[5] Of principal interest has been multilayer insulation material which consists of multiple layers of aluminized plastic films. Arcs from one aluminized layer to the next can constitute strong sources of electromagnetic interference (EMI).

Another mechanism for generating discharges is based on the fact that electrons in the energy range of 1-40 kev can penetrate materials to depths of roughly 1 μm for each 10 kev of energy. Therefore, it may be possible for electrons to be trapped below the surface of insulators causing an electric field to build up in the material. When the dielectric strength of the materials is exceeded, breakdown will occur. The resulting surface discharge can generate EMI and can damage the surface of the material. It is believed that discharges of this type have been observed in laboratory experiments.[6] Very early investigations of surface discharge phenomena were conducted by Malter.[7]

The voltage to which a surface can be charged relative to the surrounding plasma is approximately 3.6 kT_e for a Maxwellian hydrogen ion plasma of temperature T_e.[8] Although the plasma at synchronous orbit is probably by no means Maxwellian, it is reasonable to expect that the sheath potential between the plasma and shadowed surfaces of the satellite is proportional to the characteristic energy of the electron energy spectrum. There is evidence that during times of geomagnetic activity, energetic electrons are injected into the region of synchronous altitude near the midnight meridian. The motion of the electrons in the magnetic field is such that they drift from the midnight meridian towards the dawn meridian. It is believed that the charging effects observed on satellites are caused by these electrons, which are accelerated as a result of geomagnetic activity, are injected near the midnight meridian, and then drift around the Earth from midnight to dawn.

Figure 3 illustrates the geometry of this process. The geomagnetic field at high altitude is shown in the midnight meridian being swept back into a long tail as a result of interaction between the solar wind (not shown) and the geomagnetic field. During times of geomagnetic activity, it is known that the geomagnetic tail collapses back into a more dipolar configuration. Electrons trapped on these field lines are accelerated during the compression process. Kaufmann has analyzed the accelerating process and has concluded that increases in electron energy by a factor of three would be reasonable.[9] Furthermore, the flux of electrons to the satellites will scale exponentailly with the energy increase if the distribution is Maxwellian.

To summarize, there is evidence that satellites at synchronous orbit can be charged to large voltages. On the basis of this information, it has been assumed that the unexplained upset of electronic circuits on synchronous satellites is caused by EMI from arcs induced when the satellites become charged. There has not been a direct observation of circuit upset coincident with satellite-charging. However, the dependence of circuit upset upon geophysical parameters is consistent with the charging thesis.

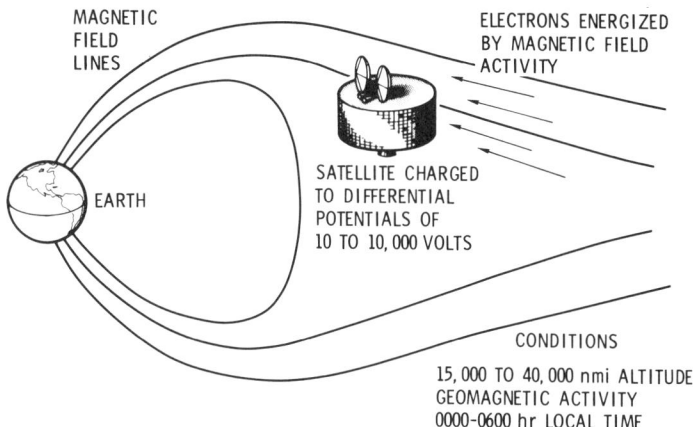

Fig. 3 Pictorial view of spacecraft-charging in the high-altitude magnetosphere.

To determine electrical charging of a satellite, special instrumentation is required, such as DeForest's electron spectrometer on ATS-5 or ATS-6. Normally, these types of instruments are carried only on special purpose scientific satellites. With the exception of an electrostatic discharge detector flown on a DSP satellite, there have been no measurements of surface charging and arcing on satellites.[10-12] At the present time, it cannot be demonstrated that spacecraft-charging is the cause of the unexplained behavior observed in so many DoD and commercial satellites. However, the evidence is fairly conclusive that spacecraft-charging can be the cause of the problem, but very little is known about the details of the process.

IV. Spacecraft-Charging Experiment

An integrated satellite experiment has been formulated: to measure the characteristics of the spacecraft-charging phenomenon, to determine the response of the satellite to the charging process, and to evaluate corrective techniques. A list of payload instruments and a brief description of objectives are provided in Table 2.

Figure 4 is a conceptual view of the satellite.[13] It is basically a right-circular cylinder 2 m diam and 2 m high with a payload weight of 86 kg. The satellite will be placed in near geosynchronous orbit (with 42,102 km apogee, 29,373 km perigee, and 2.5 deg inclination). There will be no stationkeeping and the satellite will drift at a rate of less than 4 deg/day. This is desirable because data from vehicles 3 and 4 of the DSCS II program indicate that there may be longitudi-

Table 2. Spacecraft charging instrumentation

Instrument	Objective	Experimenter(s)
Spacecraft surface-potential monitor[a]	Surface potential of 20 types of materials	H.C. Koons The Aerospace Corp.
Charging electrical effects analyzer[a]	Electromagnetic interference, 100 to 10^7 Hz	H.C. Koons The Aerospace Corp.
Quartz crystal microbalance in retarding potential analyzer[a]	Contamination rate dependence on surface potential	W.L. Lehn Air Force Materials Lab D.F. Hall The Aerospace Corp.
Thermal control sample monitor[a]	Performance of 8 thermal control materials	W.L. Lehn Air Force Materials Lab
Electric field detector[b]	Satellite potential with 50-m-long boom	T.L. Aggson Goddard Space Flight Center
Spacecraft sheath fields detector[a]	Potential measurement at three positions in the sheath	J.F. Fennel The Aerospace Corp.
Electron gun, ion gun[a]	Control of satellite potential	H.A. Cohen Air Force Cambridge Research Labs
Magnetic field monitor[b]	Triaxial measurement of geomagnetic field	D.B.G. Ledley Goddard Space Flight Center

Table 2. Spacecraft charging instrumentation (continued)

Instrument	Objective	Experimenter(s)
Light ion mass spectrometer[b]	Ion density, temperature, and drift	C.R. Chappell and D.L. Reasoner, Marshall Space Flight Center
Plasma probe[a]	Electron density and temperature	R.C. Sagalyn, Air Force Cambridge Research Labs
San Diego particle detector[b]	Electron and ion differential flux, energy, and angle resolution	S.E. DeForest, University of California, San Diego
Rapid-scan particle detector[a]	Electron and ion differential flux, time resolution	A. Pavel and L. Katz, Air Force Cambridge Research Labs
Energetic ion spectrometer[b]	1-150 AMU ion flux, 100-20,000 ev	R.L. Johnson, Lockheed Palo Alto Research Lab
Energetic proton detector[a]	Proton flux, 20-1000 kev	J.B. Blake, The Aerospace Corp.
High-energy particle detector[b]	Electron flux, 0.3-2.1 Mev, Proton flux, 1-100 Mev	J.B. Reagan, Lockheed Palo Alto Research Lab

[a] Sponsor: Air Force Systems Command.
[b] Sponsor: Office of Naval Research.

nal dependence of the anomalous behavior. The satellite will spin about the cylinder axis at a rate of 1 rpm. The desired direction of the spin vector is in the equatorial plane of the Earth and normal to the Earth-sun line. This orientation maximizes utility of data from the particle detectors because it permits detection at all angles relative to the geomagnetic field. A brief description of each instrument will be provided.

Spacecraft Surface Potential Monitor

This instrument will measure the surface potential of twenty different types of materials relative to some common reference point on the satellite. The reference point could be established by a conducting band about the satellite. Photoelectric emission will tend to keep the reference point at a small positive potential, relative to the plasma.

Figure 5 shows a concept for measuring the surface potential of an insulating material. The sample is mounted on one surface of a dielectric slab, and a conducting plate is mounted on the other surface. The capacitance of this configuration will be approximately 250 µµfarad. The conducting plate is attached to the reference point through a 0.25-µfarad capacitor. The two capacitors constitute a 1000:1 voltage divider between the sensor surface and the reference point. The principal contributor to the leakage resistance from the 0.25 µfarad capacitor is a switch (not shown) between the capacitor and the operational amplifier. This configuration can measure potential variations of the surfaces with time constants of less than 1 hr. This will be adequate for the surfaces which are rotated in and out of sunlight. Insulating surface samples in continuous shadow could be charged for long periods of time, which would introduce an error into the measurement. This problem will not exist for potential measurements on conducting surfaces.

Charging Electrical Effects Analyzer

This instrument will measure EMI in the frequency range 100-10^7 Hz. Three separate instruments will be used. A swept frequency analyzer will monitor the frequency band 0.1-10 MHz. The frequency band 100-50 kHz will be monitored by a 10-channel, fixed-frequency analyzer. There also will be capability to telemeter broadband, undetected signals from sensors in the frequency band 100-5000 Hz.

The analyzer will sample signals from a variety of sensors. These are: 1) solar array bus, 2) power line bus, 3)

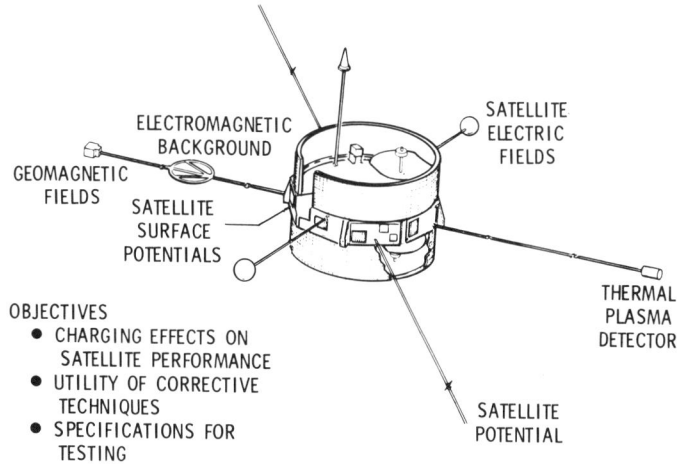

Fig. 4 Concept for instrument layout on SCATHA satellite.

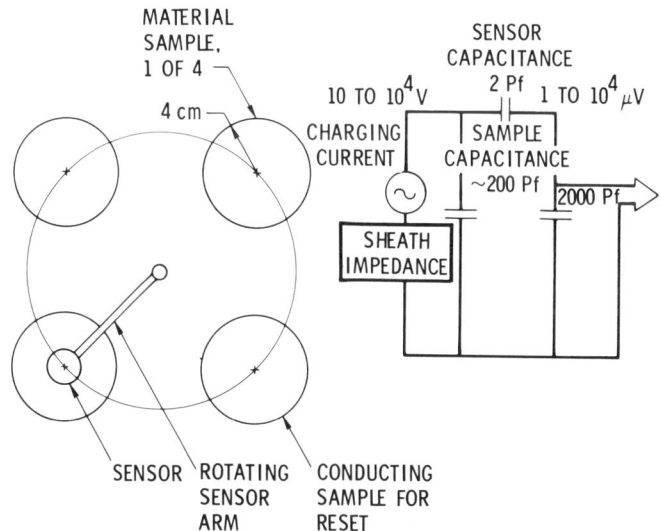

Fig. 5 Technique for measuring surface potentials on insulating surfaces.

typical command line, 4) internal loop, 5) external loop, 6) external short dipole, and 7) electric field detector boom. These sensors will permit measurement of radiated and conducted EMI, both external and internal, to the satellite.

Quartz Crystal Microbalance

Two quartz crystal microbalances placed in retarding potential analyzers will be used on the satellite. One instrument will be on the side of the satellite, and one will be placed on the end and maintained in continuous shadow. The basic instrument is similar to that used on OGO-6.[14] The instruments will have active temperature control so that the quartz sensors can be operated over a range of temperatures from -60° to 100° C. The microbalances will be placed in retarding potential analyzer structures as shown in Fig. 6. The objective of this configuration is to determine the dependence of contamination rate upon surface potential.

Thermal Control Sample Monitor

This instrument measures the backface temperature of eight thermal control material samples.[15] These instruments will be positioned contiguous with the quartz crystal monitors. Provision has been made to heat the samples and to purge contaminants that freeze out on the test surfaces. The objective of the experiment is to evaluate performance of thermal control materials as a function of on-orbit contamination conditions.

Electric Field Detector

The sensor for this instrument is a 100-m tip-to-tip dipole antenna. The antenna elements are copper-beryllium STEM extendible antennas, which are 0.64-cm-diam tubes when extended. The antenna elements will be insulated except for a few meters at the end. The absolute potential between the

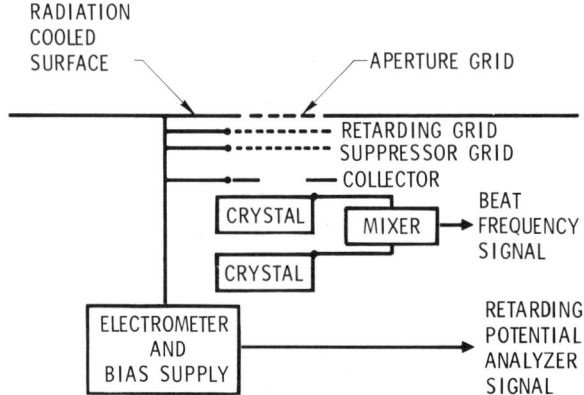

Fig. 6 Combined quartz crystal microbalance and retarding potential analyzer.

satellite and the plasma will be measured with these booms. The boom tip provides a relatively simple geometry to model for predicting the potential drop between boom and undisturbed plasma. The 50-m-length of the booms should be sufficient to extend beyond space vehicle associated sheath fields for most conditions.

Spacecraft Sheath Fields Detector

This instrument consists of three electrostatic analyzers. Two analyzers are mounted in the diametrically opposed spheres denoted as "satellite electric fields" in Fig. 4. The third analyzer is mounted on the body of the satellite. The three instruments have the same look-direction, so that if there were not electric fields about the satellite, all three instruments would record the same flux, spectrum, and angular distribution of electrons and protons in the energy range 1-1000 ev. Because there is a potential distribution about the satellite, the analyzers will detect three different spectra, each shifted according to the potential at each analyzer.

Particle Gun

The electron gun will be used to control the satellite potential for instrument calibration and to evaluate the efficacy of an electron gun to reduce differential charging of a satellite. It has been shown on ATS-5 and ATS-6 that the electron neutralizer of a cesium-ion thruster is very effective in controlling the absolute potential of a satellite.[16]

The electron gun will supply currents between .01 and 30 ma at acceleration voltages between 0 and 3 kev. Beam focussing can be varied from none to maximum capability. An argon-ion gun also is planned for the spacecraft-charging experiment. The characteristics will be 1 to 3-ma beam current at 2-kev accelerating potential.

Particle and Fields Instrumentation

The remainder of the instrumentation is typical of the particle detectors and magnetometers used on previous scientific satellites. The San Diego particle detector is the same instrument used by DeForest on ATS-5 and ATS-6 to detect spacecraft-charging. This instrument measures energy spectra in 64 steps between 1 and 70,000 ev. The acceptance angle of the telescope is 5° half-angle and the energy spectra will be measured in only four steps. Because of the high detector counting rates, standard channeltron multipliers probably would not survive. Other detectors are under consideration.

The light-ion spectrometer is basically the same instrument as flown on OGO-5.[17] One additional sensor will be added, and retarding potential grids will be incorporated so that plasma drift can be measured. The plasma probe is a spherical electrostatic analyzer using an electrometer for a detector.[18] The measurement of prime interest is electron temperature. The difficulty of this measurement is contamination of the sensor signal by photoelectrons and secondary electrons from the satellite.

The particle flux, energy spectra, and angular distribution measurements from the light-ion spectrometer, San Diego particle detector, rapid-scan particle detector, and plasma probe are considered to be critical to the success of the spacecraft-charging experiment. It is the particles in the energy range 1-50,000 ev that are believed to control spacecraft charging. Figure 7 shows the energy ranges for which the various particle detectors are designed.

V. Conclusion

The unexplained behavior exhibited by synchronous-altitude satellites is not a common every-day experience. A given satellite may experiemce no more than 10-20 circuit upsets in a year. It has been necessary to accumulate years of orbital data in order to demonstrate conclusively that the unexplained or spurious behavior depends upon geophysical parameters such as local time and geomagnetic activity. This evidence, plus the independent observations of DeForest on ATS-5, have led to the thesis that the anomalous behavior of satellites may be a result of satellite-charging by the space environment. There has been no direct observation that this is the case.

Because the problem exists today, it is necessary to assume that the spacecraft-charging thesis is correct and to implement corrective measures. It is not known whether the corrective measures, such as grounding thermal insulation blankets, are valid. Furthermore, it is not possible to test the utility of the corrective measures, because there are insufficient data regarding the response of the satellite to the space enviornment. The spacecraft-charging experiment has been designed to obtain the data necessary to specify design criteria and test procedures to insure that satellite operation is not degraded by charging effects.

In addition to the spacecraft-charging experiment, NASA Lewis Research Center plans a complementary laboratory test program to be conducted before the satellite launch in 1978. The objectives of this program are to determine the basic

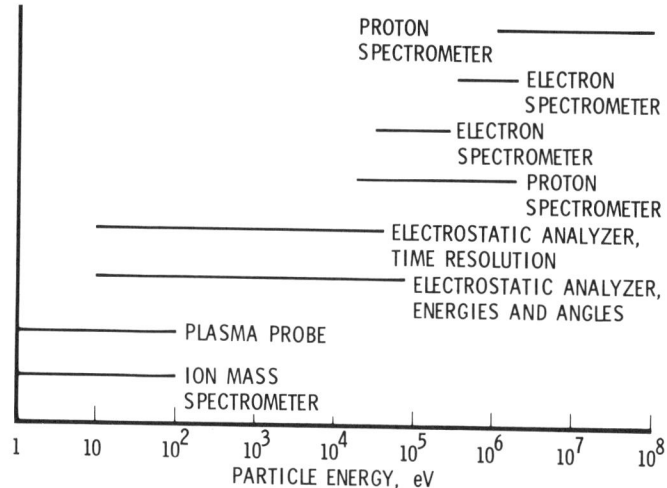

Fig. 7 Energy spectra detected by SCATHA particle detectors.

phenomenology of electrostatic charging of materials by the space environment.

References

[1] DeForest, S.E., "Spacecraft Charging at Synchronous Orbit," Journal of Geophysical Research, Vol. 77, Feb. 1972, pp. 651-659.

[2] Cauffman, D.P. and Maynard, N.C., "Model of the Effect of the Satellite Photosheath on a Double Floating Probe System," Journal of Geophysical Research, Vol. 79, June 1, 1974, pp. 2427-2438.

[3] Chappell, C.R., Harris, K.K., and Sharp, G.W., "A Study of the Influence of Magnetic Activity on the Location of the Plasmapause as Measured by OGO-5," Journal of Geophysical Research, Vol. 75, Jan. 1, 1970, pp. 50-56.

[4] Montgomery, M.D., Asbridge, J.R., Bane, S.J., and Hones, E. W., Jr., "Low Energy Electron Measurements and Spacecraft Potential: Vela 5 and Vela 6," Photon and Particle Interactions with Surfaces in Space, D. Reidel Pub. Co., Dordrecht-Holland, 1973.

[5] Fredricks, R.W. and Scarf, F.L., "Observations of Spacecraft Charging Effects in Energetic Plasma Regions," Photon and Particle Interactions with Surfaces in Space, D. Reidel Pub. Co., Dordrecht-Holland, 1973.

[6] Balmain, K.G., "Charging of Spacecraft Materials Simulated in a Scanning Electron Microscope," *Electronics Letters*, Vol. 9, Nov. 15, 1973, pp. 544-546.

[7] Malter, L., "Anomalous Secondary Electron Emission," *Physical Review*, Vol. 49, June 1, 1936, p. 879.

[8] Self, S.A., "Exact Solution of the Collisionless Plasma-Sheath Equation," *Physics of Fluids*, Vol. 6, Dec. 1963, pp. 1762-1768.

[9] Kaufmann, R., "Electron Acceleration during Tail Collapse," *Journal of Geophysical Research*, Vol. 79, Feb. 1, 1974, pp. 549-553.

[10] Nanevicz, J.R., Adams, R., and Scharfman, W., "Satellite Lifetime Monitoring," SRI Project No. 2611, March 1974, Stanford Research Institute, Menlo Park, Calif.

[11] Cauffman, D.P. and Shaw, R.R., "Transient Currents Generated by Electrical Discharges," accepted for publication in *Space Science Instrumentation*.

[12] Shaw, R.R., "Geomagnetic Substorm Activity Study Final Report," 5120, Feb. 1975, Aerojet ElectroSystems Co., Azusa, Calif.

[13] "777 Secondary Space Vehicle Conceptual Design and Flight Feasibility Study," 26598-6001-RU-00, Vol. 1, July 1974, TRW Systems Group, Redondo Beach, Calif.

[14] McKeown, D. and Dummer, R.S., "Gas-Surface Energy Transfer Experiment for OGO-F," *IEEE Transactions on Geoscience*, Vol. GE-7(2), April 1969, pp. 98-106.

[15] Luedke, E.E. and Kelley, L.R., "Development of Flight Units for Thermal Control Coatings Space Experiment," AFML-TR-72-233, Oct. 1972, Air Force Materials Lab., Wright-Patterson AFB, Ohio.

[16] DeForest, S.W. and Goldstein, R., "Study of Electrostatic Charging of ATS-5 Satellite during Ion Thruster Operation, Final Technical Report," Contract NAS JPL 953675, Dec. 1973, Jet Propulsion Lab., Pasadena, Calif.

[17] Harris, K.K. and Sharp, G.W., "OGO V Ion Spectrometer," *IEEE Transactions on Geoscience*, Vol. GE-7(2), April 1969, pp. 93-98.

[18] Sagalyn, R.C. and Smiddy, M., "Charged Particle Measurements by Means of Electrostatic Probes," *Cospar Techniques Manual Series--Electron Density and Temperature Measurements in the Ionosphere*, Cospar, Paris, France, 1967, pp. 90-111.

A HIGH-TIME RESOLUTION SPECTROMETER FOR 0.05 TO 500 keV ELECTRONS AND PROTONS

B. Sellers,[*] F. A. Hanser,[†] P. R. Morel,[≠] and
J. L. Hunerwadel,[§]

Panametrics, Inc., Waltham, Mass.

and

A. L. Pavel,[¶] L. Katz,[**] and P. L. Rothwell,[††]

Air Force Geophysics Laboratory, Hanscom Air Force Base, Bedford, Mass.

Abstract

An electron-proton spectrometer has been designed for use on the synchronous orbit research satellite, SCATHA. The spectrometer covers the energy range from 0.05 keV to

---

Presented as Paper SA61 of the SPR-Aeronomy Special Session on Spacecraft Charging by Magnetospheric Plasma, cosponsored by AIAA, at the 56th Annual Meeting of the American Geophysical Union, Washington, D. C., June 16-19, 1975. Work performed partially under Contract F19628-74-C-0217 issued by the Air Force Geophysics Laboratory (formerly Air Force Cambridge Research Laboratories) and supported by the Air Force In-House Laboratory Independent Research Fund.
[*]Head, Radiation Physics Dept.
[†]Head, Research Branch, Radiation Physics Dept.
[≠]Head, Systems Branch, Radiation Physics Dept.
[§]Senior Engineer, Systems Branch, Radiation Physics Dept.
[¶]Staff Physicist, Energetic Particles Branch.
[††]Chief, Energetic Particles Branch.
[**]Staff Physicist, Energetic Particles Branch.

more than 500 keV, and in the normal operating mode can make a complete spectral scan in 1 sec. The spectrometer has been designed to measure substorm particle fluxes at synchronous orbit under spacecraft-charging conditions, with the high-time resolution providing 1 sec spectral resolution and the capability for submillisecond resolution of selected energy channels. Data on the design and calibration of the spectrometer are given.

Spectrometer Design

The spectrometer package contains two complete sets of electron and proton detectors, oriented to measure particles arriving parallel and perpendicular to the satellite spin axis. A typical satellite mounting configuration is shown in Fig. 1. Each set of electron and proton detectors consists of four separate units: an electron solid state spectrometer (e SSS), a proton solid state spectrometer (p SSS), a low-energy electrostatic analyzer (LE ESA) for electrons and protons, and a high-energy electrostatic analyzer (HE ESA) for electrons and protons.

A simplified spectrometer block diagram is shown in Fig. 2, and a summary of the characteristics is given in Table 1. Note that each ESA has two outputs, one for electrons and one for protons. Each detector in Fig. 2 is designed to measure a portion of the electron and/or proton spectrum in five parts. Under normal operation, such a measurement is made every 0.2 sec so that a complete spectrum is measured every second. This is accomplished by transfer of the output of all compression counters into a single 200 bit scaler that is read out to telemetry 5 times per second. Provision is included to allow the ESA's and SSS's to be fixed independently at any energy bin by ground command and so provide a 0.2-sec time resolution measurement at the selected energies. Any one of the detectors also can be switched onto a high-frequency pulse amplitude modulated (PAM) channel through a 126-μsec digital ratemeter, and so provide even higher-time resolution at a single energy, either fixed or changing. Since the Spiraltron detectors used in the ESA's are subject to severe gain loss after high (about 10^{11}) total accumulated counts, provision also is included to turn

HIGH-TIME RESOLUTION SPECTROMETER

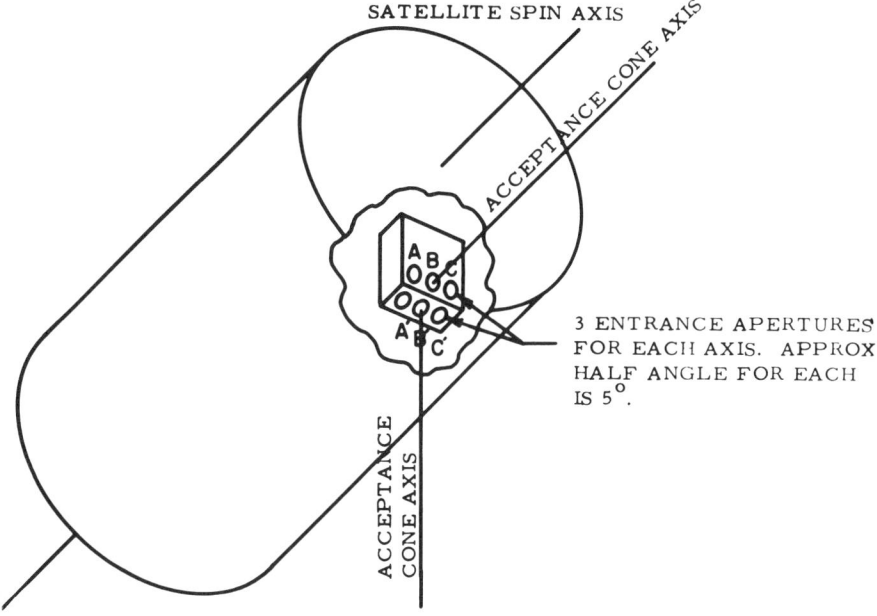

APERTURE A, A': SOLID STATE SPECTROMETERS - e
B, B': SOLID STATE SPECTROMETERS - p
C, C': ELECTROSTATIC ANALYZERS - e AND p

Fig. 1   Typical satellite mounting configuration.

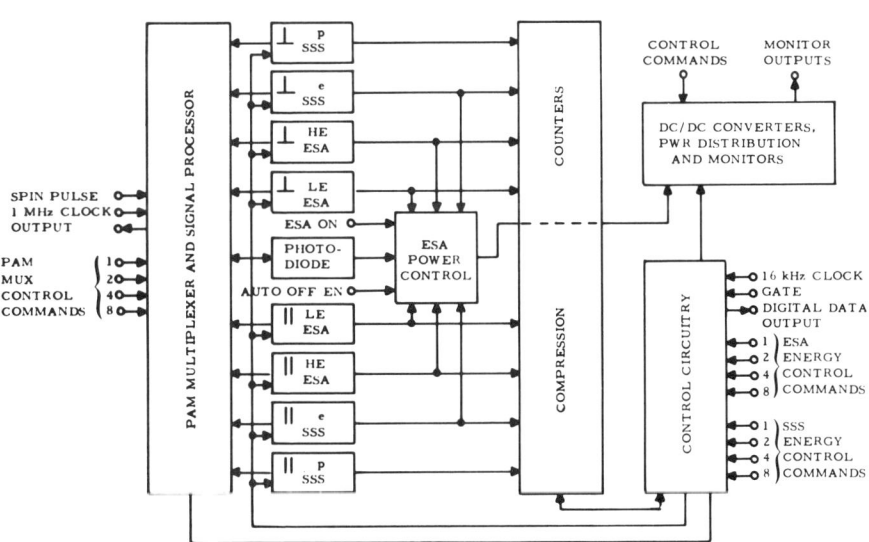

Fig. 2   Simplified instrument block diagram.

Table 1 Summary of instrument characteristics

Physical

Size	11" x 10" x 6"
Weight	13 lbs

Electrical

Input Power		5 W at 28 V ± 4 V
Commands	No. Req'd	Function
	1	Low voltage on
	1	Solid state spectr. (SSS) on
	1	Electrostatic analyzer (ESA) on
	1	ESA auto. shutoff enable
	1	Power off
	2	Spiraltron bias level control
	4	SSS energy channel selection
	4	ESA energy channel selection
	4	PAM signal selection
	1	Configuration preset

Output to Telemetry

Digital Ratemeter

Number	1
Signal Level	0-5 volts
Source Impedance	100 Ω
Format	Analog, internally subcommutated

Scaler

Number	1
Signal Level	CMOS compatible
Source Impedance	CMOS compatible
Format	Digital, 200 bits serial
Sample Rate	5 times per second

Instrument Monitors

Number	17
Signal Level	0-5 volts
Source Impedance	5 kΩ
Format	Analog, continuous
Sample Rate	Any convenient rate

off the ESA's if the count rate becomes excessive, or if a photodiode determines that one is viewing the sun. The latter provision eliminates accumulating counts from solar radiation scattered into the Spiraltrons. These automatic shutoff features can be enabled or disabled by ground command.

The solid state spectrometer designs are shown in Fig. 3. For the electron SSS, the front detector is 300 μ totally depleted with a 0.1-mil Al light and low-energy proton shield. The front detector is operated in anticoincidence with the back detector for the low energy (30 to 243-keV electrons), and in coincidence for the high-energy (243-keV to ≃ 10-MeV electrons) portion of the spectrum. The low-energy part of the electron SSS spectrum may be contaminated by 255 to 430 keV protons, but their occurrence rate normally is expected to be low compared to that for electrons, so that correction can be made from the proton SSS measurements. For the electron SSS, the front detector is 25 mm^2 in area and the back 50 mm^2. Tungsten collimators reduce the geometric

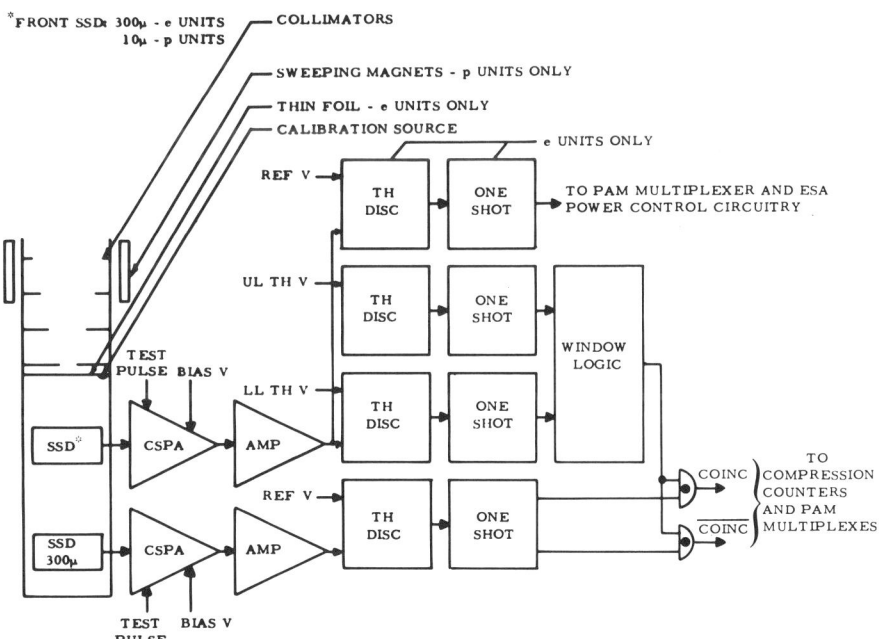

Fig. 3 Solid state spectrometers. (Abbreviations are as follows. TH DISC: threshold discriminator; CSPA: charge sensitive preamplifier; UL TH V: upper level threshold voltage; LL TH V: lower level threshold voltage.)

factor to $G_e = 3.5 \times 10^{-3}$ cm^2-sterad. A tungsten disk eliminates electrons below about 6 MeV, and protons below about 50 MeV, from rear entry to the detectors.

The proton SSS uses a 10 µ totally depleted front detector with a 120 µg/cm^2 light shield (the back ohmic contact, with the detector mounted in reverse). A magnet in front of the 10-µ detector shields it from electrons below 50 to 100 keV. The 10-µ detector in anticoincidence with the back detector measures low-energy (70 to 725 keV) protons, and in coincidence, high-energy (725 keV to 35 MeV) protons. The 10-µ detector is 10 mm^2 in area, and the back detector is 50 mm^2. Tungsten collimators give a geometric factor of $G_p = 2 \times 10^{-2}$ cm^2-sr, and the detectors are shielded from rear particle entry the same as in the electron SSS.

The SSS's detect particles in the coincidence and anticoincidence modes simultaneously in one of five energy ranges. In the normal mode of operation, they are stepped once every 0.2 sec to make a complete energy scan once every second. The energy detection ranges are listed in Table 2. For electrons, the effective highest energy is 560 keV, whereas for protons it is 35 MeV. Both SSS's have a weak alpha source to provide a check on operation during periods of low activity. Additional information on the SSS's is given in Ref. 1.

An outline of the electron-proton ESA's is shown in Fig. 4. A double-cylindrical-plate electrostatic analyzer is

Table 2 SSS energy detection ranges[a]

Energy channel code	Electron SSS			Proton SSS	
	Anticoin.		Coin.	Anticoin.	Coin.
	Electron	Proton	cont.		
0	30-45	255-270	bkgnd	70-100	15M-35M
1	45-70	270-290	bkgnd	100-165	6M-15M
2	70-120	290-330	560-10M	165-290	2.4M-6M
3	120-170	330-365	275-560	290-450	1.2M-2.4M
4	170-243	365-430	243-275	450-725	725-1.2M

[a]Energy detection ranges, keV, M = MeV.

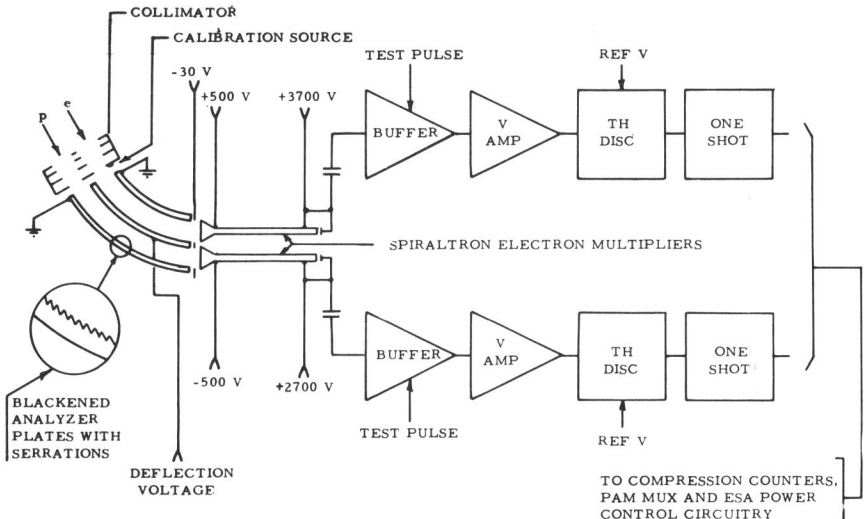

Fig. 4 Electrostatic analyzers. (Abbreviations are as follows. PAM: pulse amplitude modulated; MUX: multiplexer.)

used to select both electrons and protons with one deflection voltage. Spiraltrons are used to detect the particles. A grid at -30V before the Spiraltrons eliminates secondary electrons produced by high-energy particles striking the analyzer plates. The Spiraltron funnels are biased to preaccelerate the particles by 500V before detection. This increases the detection efficiency for low-energy particles. As shown by the electron detection efficiency measurements of Ref. 2, the preacceleration may only be necessary for electrons below 100 eV. However, there is some disagreement on the detection efficiency of channel multipliers for electron energies below 500 eV, as shown by the data of Refs. 3-5 and others cited by them, so the 500-V preacceleration avoids this region of ambiguous channel multiplier detection efficiency. For protons, the detection efficiency of channel multipliers definitely falls off rapidly below 500 eV[6-8], so the 500-V preacceleration is necessary to achieve reasonable detection efficiencies for 50-eV protons.

The analyzer plates are serrated and blackened to reduce Spiraltron response to solar radiation. A weak electron source is used to check operation of the electron ESA during quiet periods. Two ESA's are used, low-energy (0.05 to 1.7

keV) and high-energy (1.7 to 60 keV). Both measure the particles in four energy bins, with a fifth for background measurement; a complete energy scan requires one second. The low-energy ESA's have geometric factors of 1×10^{-4} cm^2-sr for electrons and 1×10^{-2} cm^2-sr for protons. The high-energy ESA's have 1×10^{-5} cm^2-sr for electrons and 1×10^{-3} cm^2-sr for protons. The ESA energy detection ranges are listed in Table 3. The ESA's have about 80% energy resolution.

The complete spectrometer makes 16 particle-energy measurements simultaneously each 0.2 sec, 8 from the set of detectors parallel to the satellite spin axis, and 8 from the perpendicular set. Each of the units shown in the block diagram of Fig. 2 provides two simultaneous outputs. Normal operation provides a complete spectral measurement every second. Several other modes of operation, including longer cycling times or fixing at a given energy, are selectable by ground command. Additional details are given in Ref. 1.

Preliminary Spiraltron Calibration

While the geometric factors of the SSS's can be calculated quite readily from the collimator dimensions and checked with measurements on calibrated particle sources, the ESA geometric factors are more difficult to determine. The transmission characteristics of the cylindrical plate ESA's are calculable, but the effect of fringing fields and

Table 3 ESA energy detection ranges[a]

Energy channel code	Low-Energy ESA	High-Energy ESA
0	Background	Background
1	0.05-0.12	1.70-4.20
2	0.12-0.30	4.20-10.2
3	0.30-0.70	10.2-25
4	0.70-1.70	25-60

[a]Energy detection ranges, keV (electron and proton).

internal scattering introduces some uncertainty into these
calculations. Finally, the efficiency of the Spiraltrons is
not unity, as is the case with solid state detectors, and varies
across the face of the funnel. Thus, the ESA's will undergo a
thorough calibration procedure using electron (beta) sources.
The entire procedure is not discussed here, but the results of
some tests on the Spiraltrons are presented.

Some tests of the relative response across the face of
the Spiraltron have been performed. These tests were undertaken because of the noticeably different profiles measured
for electrons[3-5] and protons.[6] The test setup is shown in
Fig. 5, where the method of scanning the funnel area with a
radioactive beta source (^{3}H or ^{63}Ni) is illustrated. The
first test made was to check the shape of the relative detection efficiency for the funnel grounded, and, at +500-V, the
configuration to be used in the ESA's. The results are shown

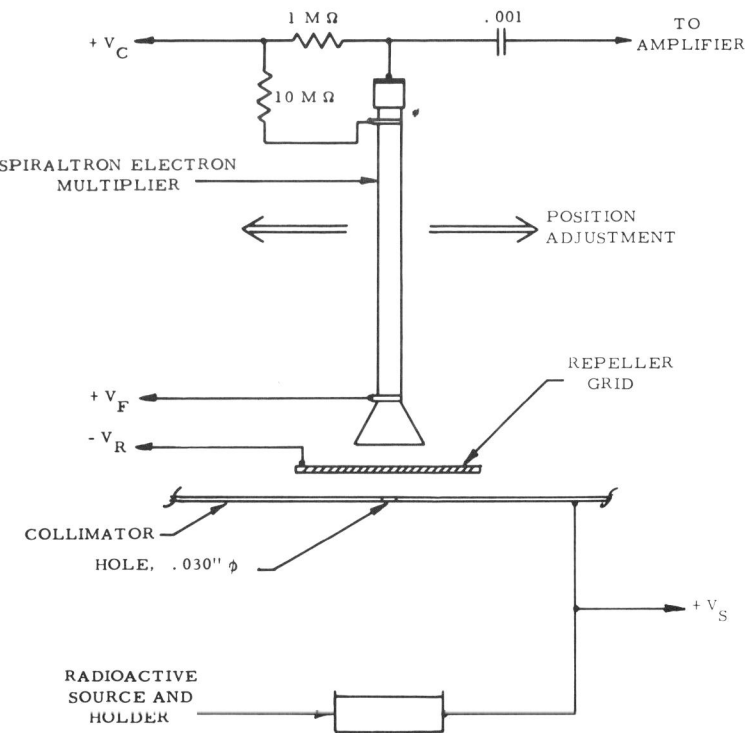

Fig. 5 Test setup for efficiency measurements of Spiraltron
electron multiplier.

in Fig. 6, where the effects of low-energy ($\lesssim 30$ eV) electrons are shown to be small by the $V_R = 0$ data being close to the $V_R = -30$-V data. The results show that detection efficiency is maximum near the center of the funnel, and falls to zero near the edge. This agrees with less detailed results reported in Refs. 3-5. We also find that applying an accelerating potential narrows the area of maximum response, although for a ^{3}H spectrum (average energy $\simeq$ 5 keV) the central response magnitude does not change very much.

Figure 7 shows the relative Spiraltron efficiency profile for three different energy ranges, with a 500-V preacceleration potential on the funnel. For the uncovered ^{3}H source with 129 μg/cm^2 of polypropylene it is about 4 keV, and for the ^{63}Ni source it is about 17 keV. Within the position accuracy shown in Fig. 7, there is no significant change in shape of the Spiraltron response. This is in agreement with the findings of Ref. 3. More refined tests of the type shown in Figs. 6 and 7 are planned. In particular, a series of tests

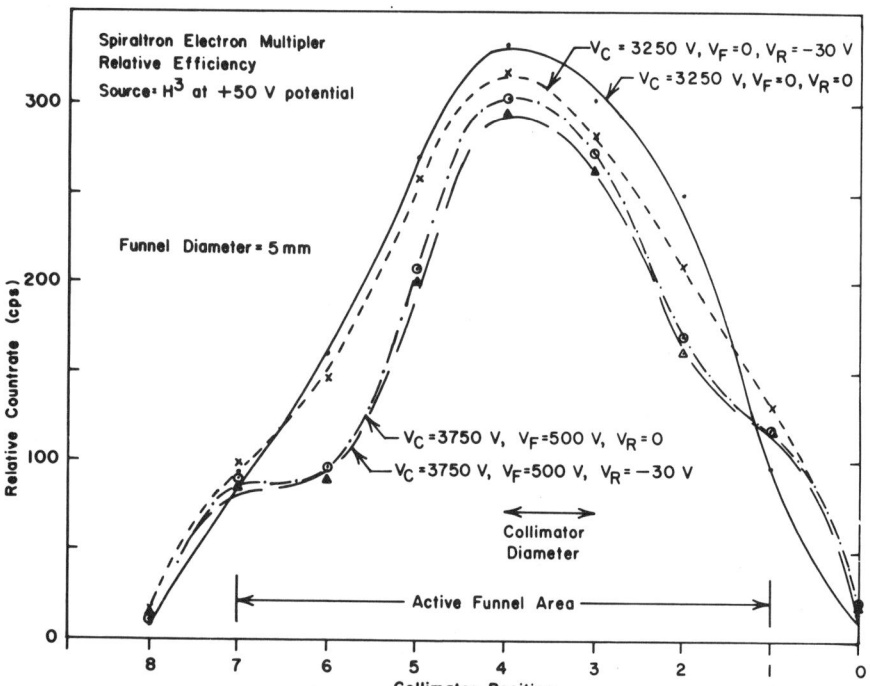

Fig. 6 Spiraltron electron multiplier relative efficiency-variation with funnel voltage. Collimator diameter shown is 0.03 in.

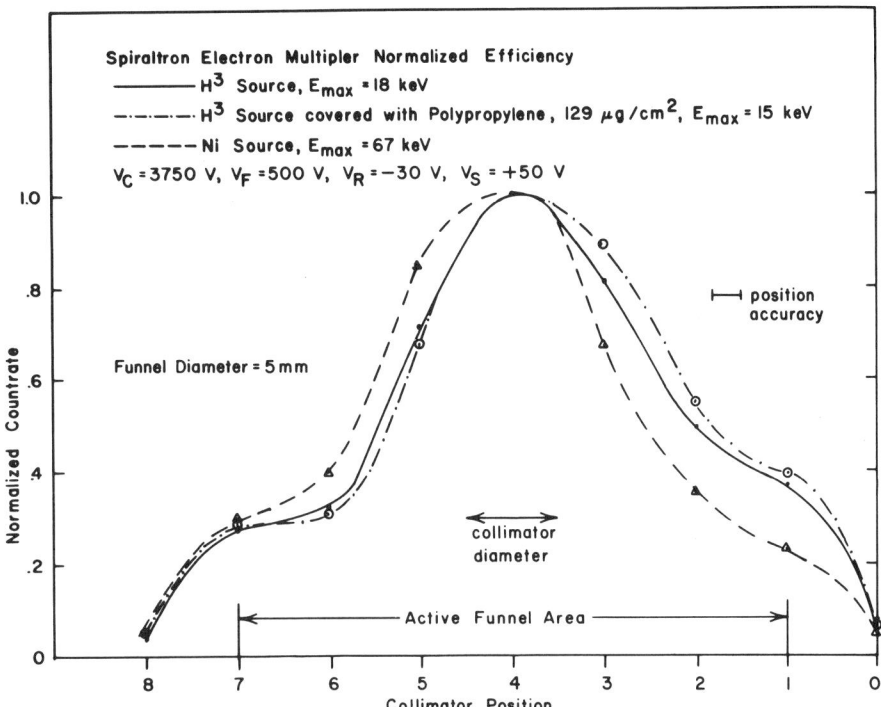

Fig. 7 Spiraltron electron multiplier normalized efficiency-variation with electron energy. Collimator diameter shown is 0.03 in.

using monoenergetic electrons from a cylindrical-plate analyzer will be carried out.

Conclusion

The design of a 0.05 to at least 500-keV spectrometer for electrons and protons has been completed. The spectrometer uses two sets of four detectors to give two view directions, parallel and perpendicular to the satellite spin axis. In order to investigate rapidly changing phenomena, the spectrometer is designed for high-time resolution, giving a complete spectral scan in 1 sec. This design complements the capabilities of the other higher-time and energy resolution instruments included in the SCATHA payload. Ground command controlled options include longer scan times and selection of a fixed set of energy bins for continuous monitoring. Submillisecond resolution on selected energy channels is possible by monitoring any one detector output on a high-frequency ($\simeq$ 10-kHz response) PAM channel.

References

[1] Hunerwadel, J. L., Morel, P. R., Hanser, F. A., and Sellers, B., "Design of Instrumentation Suitable for the Investigation of Charge Buildup Phenomena at Synchronous Orbit," Air Force Cambridge Research Laboratories, Bedford, Mass., Rept. TR-75-0365, July 1975.

[2] Bordoni, F., "Channel Electron Multiplier Efficiency for 10-1000 eV Electrons," Nuclear Instruments and Methods, Vol. 97, Dec. 1971, pp. 405-408.

[3] Sharber, J. L., Winningham, J. D., and Sheldon, W. R., "A Directional, Low Energy Electron Detector Employing Channel Electron Multipliers," Institute of Electrical and Electronics Engineers, Inc. Transactions on Nuclear Science, Vol. NS-15, June 1968, pp. 536-540.

[4] Paschman, G., Shelley, E. G., Chappell, C. R., Sharp, R. D., and Smith, L. F., "Absolute Efficiency Measurements for Channel Electron Multipliers Utilizing a Unique Electron Source," Review of Scientific Instruments, Vol. 41, Dec. 1970, pp. 1706-1711.

[5] Arnoldy, R. L., Isaacson, P. O., Gats, D. F., and Choy, L. W., "The Calibration of Electrostatic Analyzers and Channel Electron Multipliers Using Laboratory Simulated Omnidirectional Electron Beams," Review of Scientific Instruments, Vol. 44, Feb. 1973, pp. 172-177.

[6] Crandall, D. H. and Ray, J. A., "Channeltron Efficiency for Counting of H^+ and H^- at Low Energy," Review of Scientific Instruments, Vol. 46, May 1975, pp. 562-564.

[7] Iglesias, G. E. and McGarity, J. O., "Channel Electron Multiplier Efficiency for Protons of 0.2-10 keV," Review of Scientific Instruments, Vol. 42, Nov. 1971, pp. 1728-1729.

[8] Egidi, A., Marconero, R., Pizzella, G., and Sperli, F., "Channeltron Fatigue and Efficiency for Protons and Electrons," Review of Scientific Instruments, Vol. 40, Jan. 1969, pp. 88-91.

Chapter 2—Environmental Interaction with Spacecraft

Do spacecraft occasionally charge up to the multikilovolt range at synchronous altitudes? Are the observed spacecraft malfunctions related to environmental phenomena?

The papers in this section address these two questions. The first two papers report on studies that establish that many of the observed spacecraft malfunctions are related to environmental phenomena. The following papers show how the environment intereacts with a spacecraft so as to charge it up to the multikilovolt range and what active and passive techniques might be employed to control spacecraft potential.

The outstanding difficulty in performing a definitive study of substorm-related anomalies is the lack of data on the charge and discharge state of the spacecraft as a function of time. This problem is prevalent in every spacecraft system for which substorm-related events have been identified, and it accounts for the fact that it took more than two decades to find the problem. In the case of the DSCS satellite, described in the first paper in this chapter, a relatively small number of different types of spurious events were analyzed and compared with ground station data that included indications of thousands of substorm events. The results were consistent with the hypothesis that substorms may be responsible for many of the observed malfunctions. However, it has not been possible to establish a direct correspondence between anomalies and substorm phenomena in most of the malfunctions observed on operational spacecraft. The major obstacles to demonstrating a conclusive correlation are the lack of a statistically significant sample of spacecraft anomalies and/or spacecraft failures and the sparse location of ground stations for monitoring the localized environment of the spacecraft. This can be rectified by including within the onboard housekeeping system of every spacecraft a set of monitors to measure the accumulated charge and associated arc discharge at levels well below the malfunction threshold. The second paper, by Shaw, Nanevicz, and Adamo describes the utilization of such a set of monitors to verify with in situ measurements that electrical breakdowns occur on the surface of a satellite. There have been innumerable theoretical and experimental studies which show that a spacecraft immersed in a substorm plasma will charge to the multikilovolt range. Some of the latest experimental environmental data are presented in the third paper by Vogl, Sanders, and DeForest and in the fourth paper by Reasoner, Lennartsson, and Chappell. The references given in these papers provide a source of additional background environmental data. The observation of field-aligned fluxes of electrons and protons means that calculations based on isotropic distributions are not always valid. Warm plasma encounters in the noon-to-dusk local time sector also will affect the spacecraft charging and discharging characteristics, as shown in the calculations performed by G. T. Inouye. Theoretical and experimental studies of the behavior of spacecraft immersed in a plasma are presented in the last four papers in this chapter.

A CORRELATION STUDY RELATING SPACECRAFT ANOMALIES TO ENVIRONMENTAL DATA

C. P. Pike[*]

Air Force Cambridge Research Laboratories[**]
Hanscom Air Force Base, Mass.

and

M. H. Bunn[+]

U. S. Air Force Space and Missile Systems Organization
Los Angeles, Calif.

Abstract

An environmental data program was initiated to support the U. S. Air Force's Defense Satellite Communications System (DSCS) so that the operational environment for geosynchronous orbiting DSCS satellites could be specified at times of satellite anomalies. The anomalies studied included uncommanded logic reset anomalies, spinup anomalies, and power converter switching anomalies. Magnetospheric substorm activity is associated with 90% of the logic reset anomalies, with all nine of the converter switching anomalies, and with two out of three of the spinup anomalies. Magnetospheric substorms, however, occur far more frequently than anomalies occur. Although this study suggests that the environment

---

Presented as Paper SA38 at the American Geophysical Union Spring Annual Meeting (co-sponsored by the AIAA), Washington, D.C., June 16-19, 1975.
The authors gratefully acknowledge the numerous scientists who provided data.
[*]Physicist
[**]now Air Force Geophysics Laboratory
[+]Captain, U.S. Air Force

could be a factor in producing the anomalies, a conclusive argument for or against the environment acting on the DSCS satellite has not been made, mainly because of a lack of in situ environmental measurements.

Introduction

Environmental conditions prevailing during time periods in which Defense Satellite Communications System (DSCS) satellites displayed performance anomalies have been specified in an attempt to isolate environmental features or phenomena that might be a causative factor in producing the anomalies. The types of anomalies experienced by DSCS satellites include uncommanded switching of solid-state logic circuits in the satellite's communications repeater system, spinups of the satellite's despun platform, and unprogrammed switching of power converters. In this paper, results are presented of a correlation study that relates the occurrence of environmental phenomena to the occurrence of these three different types of satellite anomalies. The purpose of this study is to establish the likelihood that the environment may produce a certain type of anomaly.

Program Outline

In support of this correlation study, an extensive acquisition program was initiated for environmental data from such diverse sensors as ionosondes, magnetometers, riometers, scintillation and total electron content detectors, all-sky cameras, solar monitor stations, satellite particle flux detectors, and the U.S. Air Force's Defense Meteorological Satellite Program (DMSP) satellites. Table 1 contains a breakdown of data used and their source.

Study efforts focused on determining the state of the magnetosphere and on determining solar conditions during anomaly periods. The magnetosphere was studied because DSCS satellites are located in the magnetosphere at geosynchronous orbit, and geomagnetic field lines that intersect DSCS satellites map down to the Earth's surface near 66° corrected geomagnetic latitude[1] (CGL). (The new Olson-Pfitzer geomagnetic field model[2] would map geosynchronous orbit to $66\frac{1}{2}°$ magnetic latitude in the day sector and to $65\frac{1}{2}°$ magnetic latitude in the night sector.) In Figs 1a and 1b, the heavy dashed lines drawn on the polar geographic maps represent an approximate locus, in the northern and southern hemispheres

Table 1 Environmental data program data base

Type	Source
DMSP	Air Weather Service
Magnetometer	
Goose Bay	AFCRL
Thule	AFCRL
Churchill	National Research Council
Loring	USAF Environmental Technical
Eielson	Applications Center
Riometer	
Goose Bay	AFCRL
Great Whale	World Data Center
Churchill	Communications Research Centre
Val'dor	Communications Research Centre
Moosonee	Communications Research Centre
Ottawa	Communications Research Centre
Alaska	Air Weather Service
Scintillations	
Goose Bay	AFCRL
Hamilton	AFCRL
Narsarssuaq	AFCRL
Total Electron Content	
Goose Bay	AFCRL
Hamilton	AFCRL
Narsarssuaq	AFCRL
Ionosonde	
Churchill	Communications Research Centre
Goose Bay	AFCRL
Ottawa	Communications Research Centre
Solar Radio Network	AFCRL
Solar Optical Network	AFCRL
Particle Fluxes	
ISIS	NASA/U. Of Texas
OV5-6	AFCRL
AE-C	NASA
IMF and Solar Wind	U. Of California
Solar x-rays	Naval Research Laboratory
Vela	Los Alamos Scientific Laboratory

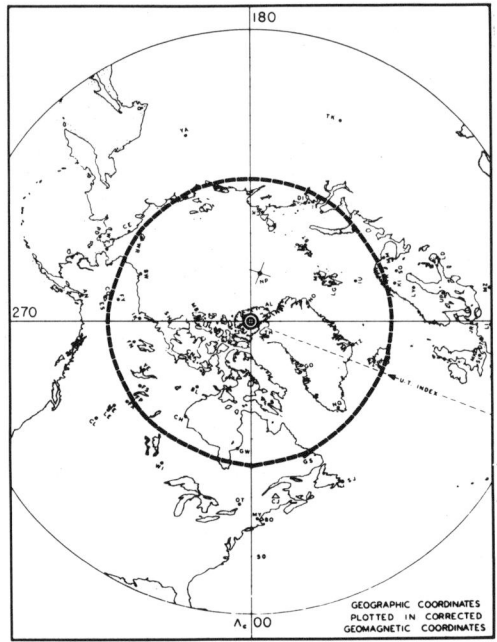

Fig. 1a Dashed line indicates a projection along magnetic field lines onto Earth's surface in the northern hemisphere of geosynchronous altitude.

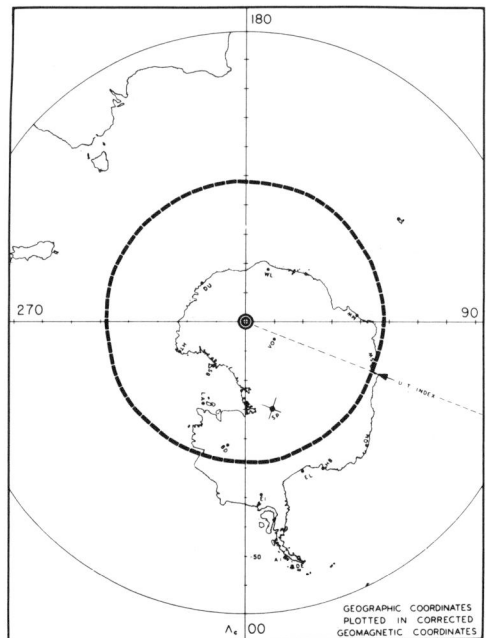

Fig. 1b Dashed line indicates a projection along magnetic field lines onto Earth's surface in the southern hemisphere of geosynchronous altitude.

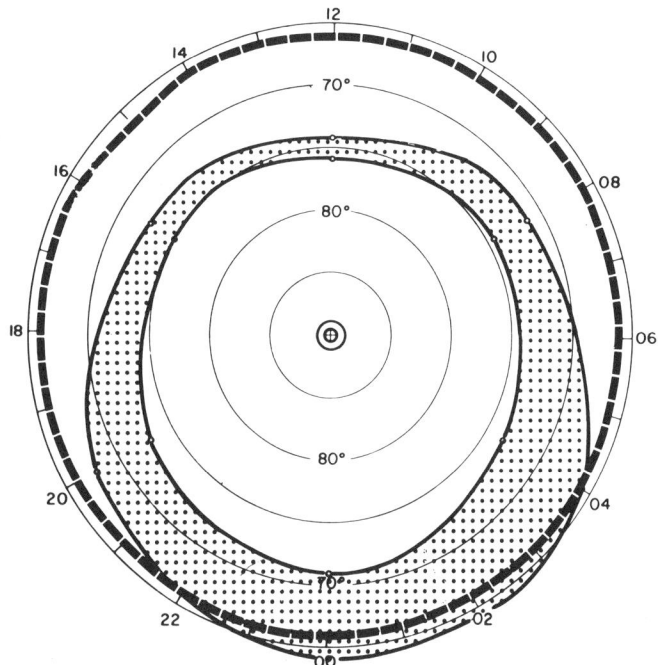

Fig. 2 Shading indicates the location of the auroral oval, drawn onto a corrected geomagnetic latitude/local time grid. Geosynchronous altitude projection onto the grid is noted by the heavy dashed line.

respectively, of the ground terminus of the geomagnetic field lines that pass through the equatorial plane at geosynchronous altitude (6.6 Earth radii). Ground-based high-latitude diagnostics such as riometers, magnetometers, ionosondes, scintillation recordings, and total electron content measurements then can be used to infer the state of the magnetosphere. Geophysical stations that are located near 66° CGL can be identified in Fig. 1, and, where possible, data from these stations were used. The photographs of auroras, recorded from DMSP satellites, are one of the best means for inferring the state of the magnetosphere. In fact, solar disturbances, which effect the magnetosphere, are reflected in the auroral photographs. The DMSP auroral photographs were, therefore, a cornerstone for the analysis and are used extensively in this paper.

The auroral oval[3], defined as the region where the probability of occurrence of auroras exceeds 0.7, is noted in Fig. 2 on a CGL and CG local time grid. The auroral oval for moderately active auroral conditions is shown by shading. In reality, auroras are a dynamic phenomena, and the auroral oval, shown here, is a statistical estimate of the location where auroras are likely to occur for a moderate level of auroral activity. By comparing the latitudes of the auroral oval to 66° CGL (noted by the heavy dashed line in Fig. 1), which corresponds to the approximate ground terminus of the geomagnetic field lines passing through the equatorial plane at 6.6 Earth radii, it is evident that, in the night sector, the plasma sheet particle fluxes that produce auroras occupy some of the same field lines that geosynchronous satellites occupy[4]. In the day sector, these same field lines typically are occupied by outer Van Allen belt particle fluxes. These fluxes also produce auroras; however, the character of the auroras and the associated energy spectrum of the particles is quite different from the character of the nightside auroras and the associated energy spectrum of the particles[5].

Correlation Study

Anomalies that occurred between November 1971 and October 1974 on DSCS Satellites 9431, 9432, 9433, and 9434 were studied. For each anomaly, data from high-latitude geophysical stations were examined in conjunction with DMSP satellite auroral photographs in an effort to infer the relative state of activity in the magnetosphere. A very limited amount of data on auroral particle fluxes were available. Solar events also were studied using data from the joint U.S. Air Force Weather Service/Air Force Cambridge Research Laboratories Solar Radio Monitoring Network and the optical observations as reported in various solar and geophysical bulletins. Although inputs from 10 different disciplines were incorporated into this study, caution should be exercised when drawing conclusions about the magnetosphere from these inputs, as no in situ measurements were available to specify the DSCS operational environment. In addition, the environmental data base is not all-encompassing; thus, some environmental factors are not explored fully.

As previously discussed, DMSP auroral photographs provided by the Air Weather Service were the cornerstone of the analysis. The photographs provided, on a rapid-response basis, a baseline value for specifying the magnetospheric disturbance level. The photographs provide graphic evidence of the occurrence of quiet periods as well as of disturbed periods in the magnetosphere. Figure 3 contains DMSP satellite photographs of the aurora. The occurrence of magnetospheric substorms at geosynchronous orbit can be monitored by identifying auroral substorm patterns in the photographs.

SPACECRAFT ANOMALIES 51

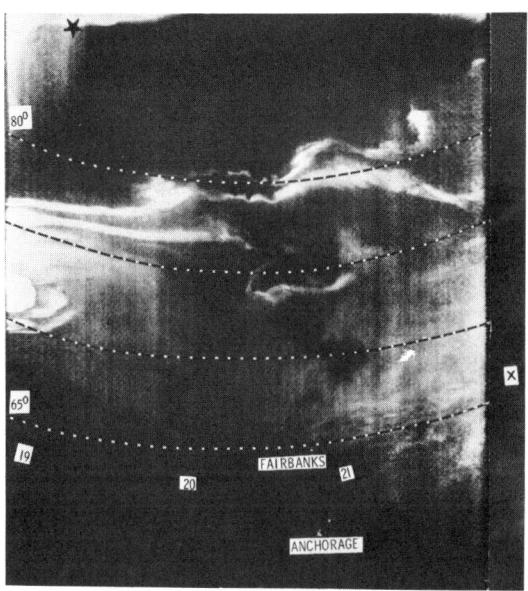

Fig. 3a DMSP auroral photograph, recorded at 0747 universal time (UT) on Nov. 25, 1971, shows auroral substorm activity. Symbol X is a ground "foot-print" of DSCS satellite located in the equatorial plane at 6.6 Earth radii. Corrected geomagnetic coordinate grid is super-imposed on the photograph.

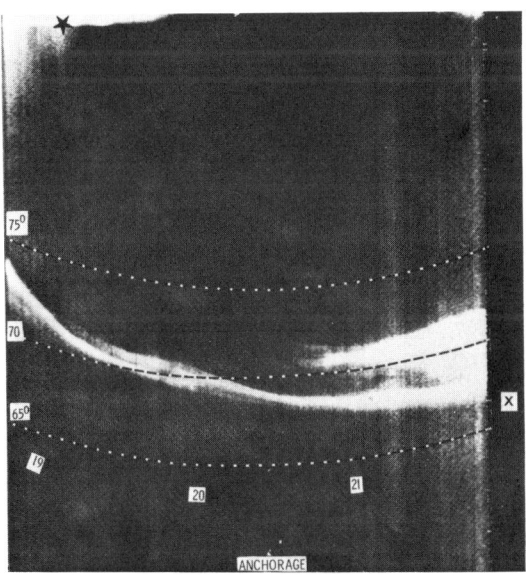

Fig. 3b DMSP auroral photograph, recorded at 0719 UT on Nov. 26, 1971, shows auroral pattern suggestive of pending auroral substorm.

That is, auroral patterns indicative of magnetospheric substorms can be seen readily in the photographs; hence, photographs recorded near times of satellite anomalies were examined for evidence of substorm occurrence to correlate with anomaly occurrence.

In general, DMSP satellites are at an altitude of about 815-850 km and are in a 99° sun-synchronous orbit. DMSP satellites are kept in noon-midnight and dawn-dusk local time orbital planes. The photographs are produced by a line scanning technique in which the field of view of the detector is swept repetitively across the Earth, perpendicular to the path of the satellite, and through the subsatellite point. The scan repetition rate and the field of view are adjusted so that the forward motion of the satellite in its orbit carries it a distance corresponding to the width of the field of view on the Earth between successive scans. The scans form a raster which composes the photographs. The orbital period of the satellite is 101 min. therefore, the time required to generate a photograph, 10° in latitudinal extent (which is a typical width of the auroral region), is 2.8 min. The field of view of the detector at the subsatellite point is 3.7 km at the surface of Earth and 3.2 km at 100-km altitude. The longitudinal width of the photographs is 3000 km at the surface of the Earth and 2500 km at 100-km altitude (the altitude of the aurora). The spectral range of the detector is 0.4 to 1.1 μ, and the detector is most sensitive near 0.8 μ. The minimum detectable auroral brightness is estimated to be International Brightness Coefficient (IBC)~I; saturation brightness is estimated to be IBC ~ 3. Auroras appear to brighten progressively from the center of the photograph to the edge of the photograph as a result of the "Van Rhijn effect": the increase in optical thickness with increasingly oblique viewing angle.

Logic Reset Anomalies

Shortly after the first two DSCS satellites were launched and commanded into operational configuration, they started experiencing uncommanded switching of solid-state logic circuitry, primarily in the tunnel diode amplifier limiter (TDAL). Although these logic resets could not be monitored on satellite 9432 after September 1972 (due to loss of the telemetry downlink) and on satellite 9431 after June 1973 (due to loss of secondary power distribution to the despun platform), design changes on satellites 9433 and 9434 were instituted. These include the addition of noise-desensitizing resistors as well as an attempt to ground electrically all external conductive surfaces in excess of 36 in^2. Nevertheless, satellites 9433 and 9434 experienced identical logic resets eight days after being placed in orbit, whereas hundreds of hours of ground testing of these systems gave no indication of this type of disturbance.

Early logic upsets occurring on the first two satellites were of such a nature that six consecutive commands would have been required to cause the transition from the nominal gain state to that existing after the anomaly, a highly unlikely mechanism. Extraneous noise introducing transients into the power distribution lines or input signal lines of the switching logic assembly (SLA), which commands TDAL gain changes, could produce the repeated upsets. (Results of the complete investigation are given in Ref. 6.) Consideration of all possible internal and external sources of such extraneous noise, including radiation from ground-based radar and other satellites as well as environmental radiation, suggests that coupling between the geosynchronous altitude environment and the satellites is such that sufficient differential voltages could be induced onto the spacecraft to cause electrical discharges of sufficient magnitude to interrupt these circuits.

Figure 3a shows the auroral signature of a magnetospheric substorm that occurred near the time of a logic reset anomaly. The ground terminus of the geomagnetic field line that intersects the equatorial plane of the Earth at the longitude of the DSCS satellite is noted by the symbol X. Figure 3a was recorded during new moon conditions and the Earth's surface, which provides the background for the picture, is black. Auroras appear in the foreground of the picture as the white region. The cities, Fairbanks and Anchorage, Alaska, have been labeled for reference. The DMSP photograph in Fig. 3b shows an auroral signature that would be the precursor of a substorm similar to the substorm seen in Fig. 3a. In fact, a logic reset anomaly occurred 1 hr after this photograph (Fig. 3b) was taken, suggesting that, at the time of the anomaly, a substorm was in progress.

In Fig. 4, a sequence of DMSP photographs recorded at 1937, 2119, and 2301 universal time (UT) on Feb. 23, 1973 documents the pronounced substorm activity occurring at this time. The heavy dashed line, marked L=6.6 and extending across each photograph, represents the projection along geomagnetic field lines to the Earth's surface of geosynchronous altitude (6.6 Earth radii). Bright auroral forms indicative of substorm activity are seen in the midnight time sector in all photographs. A logic reset anomaly occurred at 2022 UT, which is during a time period of intense substorm activity. It should be pointed out that the DSCS satellite

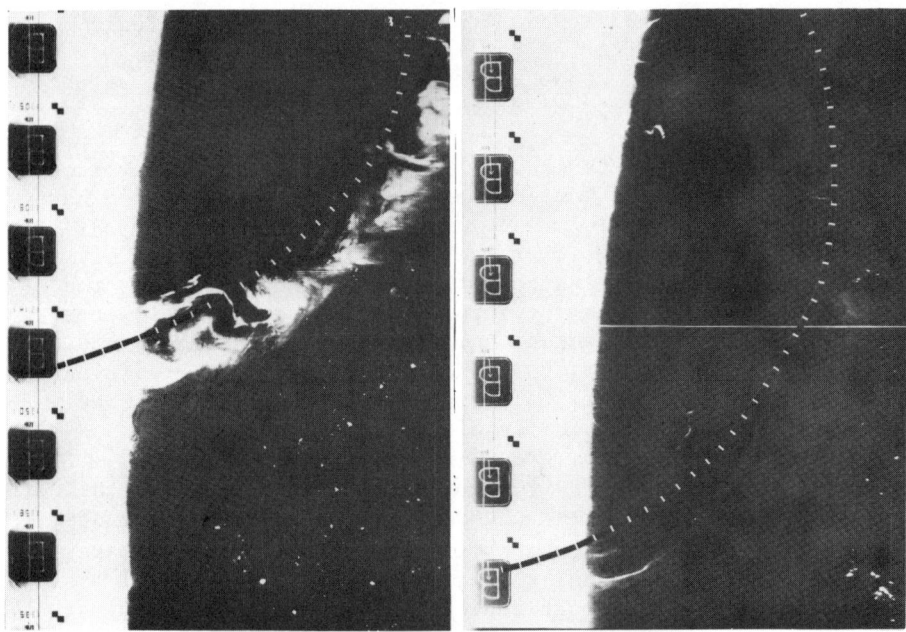
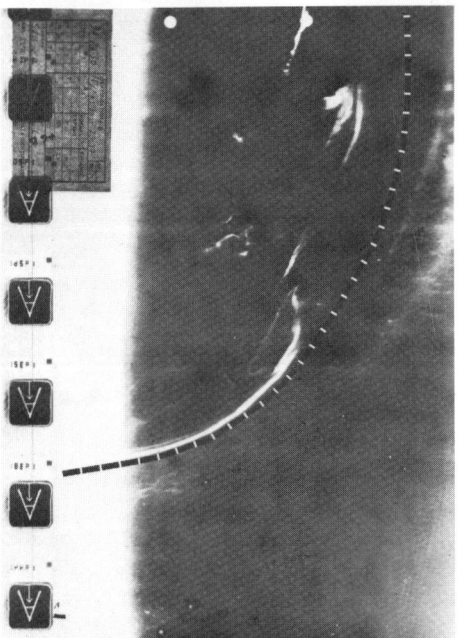

Fig. 4 DMSP auroral photographs, recorded at 1937, 2119, and 2301 UT on Feb. 23, 1973. Geosynchronous altitude projection onto Earth's surface is noted by a heavy dashed line. Auroral substorm activity is evident in the photographs.

was in the day sector at the time of the anomaly, whereas these photographs document auroral substorm activity near midnight. Substorm effects are, however, occurring in all local time sectors and, hence, at the longitude of the satellite.

Despun Platform Power Failure

On June 2, 1973 at 0737 UT, Earth coverage beacon, spin-up monitor alarms were triggered at the Satellite Test Center while monitoring a DSCS satellite. Telemetry data revealed, however, that no spin-up had occurred, but that the communications subsystems on the despun platform were not receiving electrical power. An intensive investigation into this catastrophic failure was conducted by the spacecraft prime contractor[7], which concluded that the most probable cause of this failure was an overloading of primary power circuitry due to electrostatic discharging of differentially charged surfaces on the spacecraft. In Fig. 5, a DMSP photograph recorded at 0741 UT shows that intense auroral substorm

Fig. 5   DMSP southern hemisphere auroral photograph recorded at 0741 UT June 2, 1973. An intense auroral substorm is in progress.

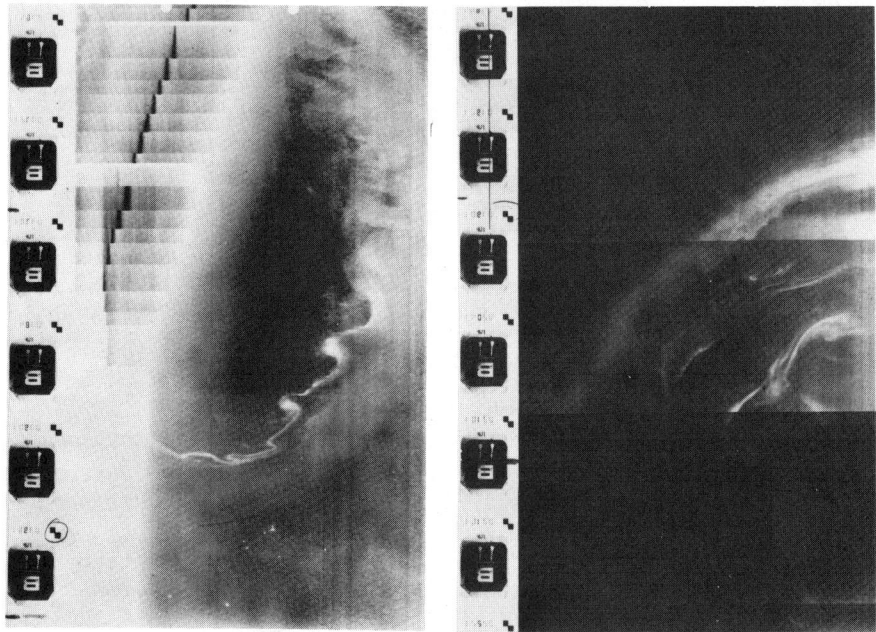
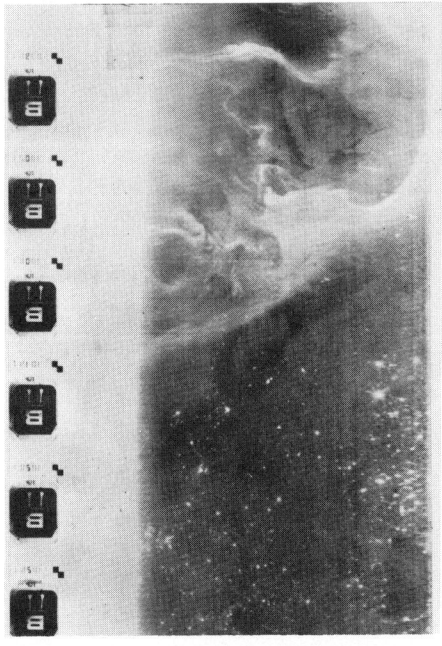

Fig. 6 DMSP auroral photographs recorded at 1511, 1830, and 2017 UT on Oct. 14, 1974. Intense auroral substorm activity is evident.

activity was in progress at this time. In general, numerous auroral photographs have been examined, and, based on this examination, it is apparent that the photograph in Fig. 5 is unique because of the high-intensity level of auroral emission evident. The emission comes from a spatially limited area and saturates the satellite sensor. This indicates that high auroral particle flux levels exist at this time. The DSCS satellite was located in the dawn sector at the time of the anomaly and would, most likely, be subject to high particle flux levels.

Logic Reset, Power Converter Switching, and Spinup Anomalies

Erroneous command executes and logic upsets in the design electronics assembly (DEA) have caused DSCS satellites to revert from the normal Earth-lock mode to a spinup condition three times in the past year. On Oct. 14, 1974 a spinup anomaly occurred at 2007 UT. On that same date, a power converter switching anomaly (discussed in more detail in a forthcoming paragraph) occurred at 1549 UT, as well as a logic reset anomaly at 1741 UT. The DMSP auroral photographs recorded on Oct. 14, 1974 at 1511, 1830, and 2017 UT are seen in Fig. 6. In general Oct. 14 was a disturbed day, and substorms occurred during the time period in which the satellite anomalies occurred. The DSCS satellite was located in the morning and prenoon time sectors at the times of anomalies.

Converter Anomaly

DSCS satellite 9434 also has exhibited another abnormal operation of the switching logic assembly (SLA). On several occasions, an SLA converter has turned off or switched back and forth to the redundant unit, causing communications outages of up to 5 hrs. These anomalies have been shown to be produced by disturbances on the secondary power lines such as have been discussed earlier, although this may not be a single feasible causative factor[8].

On Oct. 18, 1974 at 2359 UT, a power converter switching anomaly occurred, accompanied by other logic circuitry upsets. In Fig. 7, the DMSP auroral photographs recorded at 1915 and 2057 UT Oct. 18, 1974 and 0202 UT Oct. 19, 1974 show intense substorm activity occurring at the time of the switching anomaly. The DSCS satellite was located in the afternoon time sector at the time of the anomaly.

Conclusion

Of the approximately 50 logic reset anomalies examined in the manner just described, approximately 90% of the anomalies

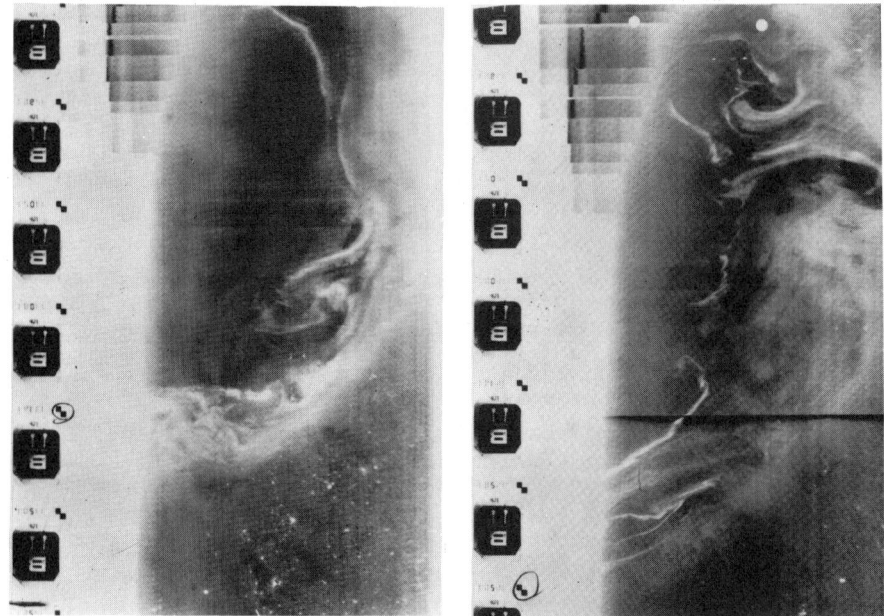
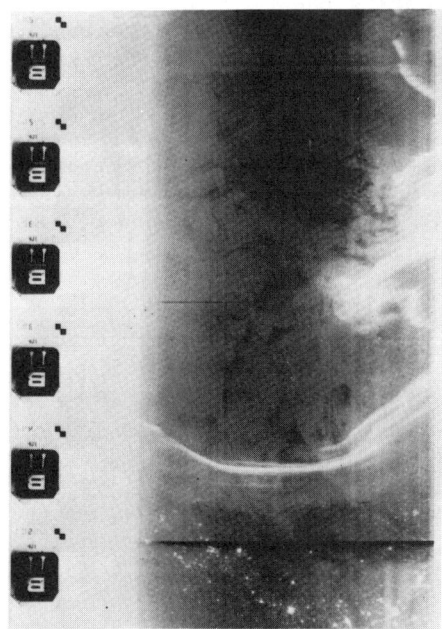

Fig. 7 DMSP auroral photographs recorded at 1915 and 2057 UT on Oct. 18 and 0202 UT on Oct. 19, 1974. Intense auroral substorm activity is evident.

have occurred at times when auroral and magnetospheric substorms also have occurred, whereas 10% of these reset anomalies appear to have occurred at times relatively free of substorms. Many logic reset anomalies are correlated closely with solar flares/solar noise storms; however, there are many periods of solar activity with no anomalies occurring[9]. A possible correlation also exists between increases in the gamma ray flux measured on the satellite OV5-6 and gain state change anomalies[10]. Six of these anomalies apparently are related closely to increases in gamma ray flux. Of the nine power converter switching anomalies that have occurred, all nine are correlated positively with substorm activity, whereas of the three spinup anomalies studied, two are correlated positively with substorm activity.

Substorms, however, are not uncommon occurrences, whereas the anomalies occur relatively infrequently as compared to the occurrence of substorms. In fact, during August 1972, the major geomagnetic storm of the present solar cycle occurred; yet no anomalies were recorded. With the present data base, it is not possible to quantize substorms, and, in fact, quantization of substorms is just at the upper limit of the present state of knowledge of the magnetosphere. This is in part due to the extreme variability of magnetospheric particle fluxes which occurs during substorms and to the problems that then are associated with typifying a substorm in terms of a relative magnetospheric particle flux level or of a charge deposition rate on a satellite. Thus, this study suggests that the substorm environment may be a factor in producing anomalies and that effects of spacecraft charging on the DSCS satellites could be an underlying causative factor in producing the anomalies.

References

[1]Hakura, Y., "Table and Maps of Geomagnetic Coordinates Corrected by the Higher Order Spherical Harmonic Terms," Report on Ionospheric and Space Research, Japan, No. 19, 1965, pp. 121-157.

[2]Olson, W.P. and Pfitzer, K.A., "A Quantitative Model of the Magnetospheric Magnetic Field," Journal of Geophysical Research, Vol. 79, No. 25, September 1974, pp. 3739-3748.

[3]Feldstein, Y.I., "Some Problems Concerning the Morphology of Auroras and Magnetic Disturbances," Geomagnetism and Aeronomy, No. 3, 1963, pp. 183-186.

[4] Akasofu, S.-I., DeForest, S., and McIlwain, C., "Auroral Displays Near the "Foot" of the Field Line of the ATS-5 Satellite," <u>Planetary Space Science</u>, Vol. 22, No. 1, January 1974, pp. 25-45.

[5] Winningham, J.D., Akasofu, S.-I., Yasuhara, F., and Heikkila, W., "Simultaneous Observations of Auroras from the South Pole Station and of Precipitating Electrons by ISIS 1," <u>Journal of Geophysical Research</u>, Vol. 78, No. 28, October 1973, pp. 6579-6594.

[6] "Final Technical Report, Program 777 Anomaly Investigation for Satellites 9433 and 9434," TRW Systems Group Technical Report, No. 09670-RFP-050-01, March 22, 1974.

[7] "Final Technical Report, Performance Anomaly Flight 9431, June 2, 1973," TRW Systems Group Technical Report No. 24512-AR-006-01, Aug. 21, 1973.

[8] "Final Technical Report, Program 777 Anomaly Investigation for Satellite 9434 SLA Convertors," TRW Systems Group Technical Report No. 26722-AR-008-01, Oct. 25, 1974.

[9] Castelli, J.P., private communication, May 1975, Air Force Cambridge Research Laboratories.

[10] Yates, G.K., private communication, May 1975, Air Force Cambridge Research Laboratories.

OBSERVATIONS OF ELECTRICAL DISCHARGES
CAUSED BY DIFFERENTIAL SATELLITE-CHARGING

R. R. Shaw*
Aerojet ElectroSystems Company, Azusa, Calif.
and
J. E. Nanevicz† and R. C. Adamo‡
Stanford Research Institute, Menlo Park, Calif.

Abstract

An experiment has been flown on a satellite at synchronous altitude to study differential electrical charging of satellite thermal-control materials and to detect the presence of dielectric breakdowns caused by such charging. Electric discharges have been observed by the experiment at essentially all points in orbit. The discharges have been categorized into two types, the first of which is strongly spin synchronous and occurs during undisturbed geomagnetic periods. The second type of discharge is observed near local midnight during magnetospheric substorms (as determined both by the presence of energetic electrons seen by the experiment and by ground-based geomagnetic indices). Discharges have been observed coincidently with two types of engineering orbital anomalies that occur on this satellite, strongly suggesting that the anomalies were caused by current surges or electromagnetic inter-

Introduction

Satellites at synchronous orbit have been observed to charge to large voltages as a result of the presence of charged particle currents associated with the spacecraft

---

Presented as Paper SA-41 at the 1975 Spring Annual Meeting of the American Geophysical Union, Washington, D. C., June 16-19, 1975.
 *Engineering Specialist; now at the Department of Physics and Astronomy, The University of Iowa, Iowa City, Iowa.
 †Program Manager
 ‡Research Engineer

ference from the discharges. In addition, a statistical correlation between anomalous temperature rise rates on these satellites and geomagnetic indices suggests the possibility that part of the degradation of the satellite thermal control surfaces may be due to physical damage from the discharges. environment. During energetic particle-injection events associated with magnetospheric substorms, ATS-5 has been observed to charge to negative voltages as great as several thousand volts when the satellite is eclipsed by Earth.[1] All satellites that operate near synchronous orbit altitudes are charged by currents collected from the plasma and by currents emitted from the surface of the satellite. The resulting surface voltages are determined by characteristics of the immediate environment, the external configuration of the satellite, and the properties of the external surface materials used on the satellite.

If a satellite has insulating outer surfaces, those surfaces can charge to large differential voltages with respect to the conducting frame (referred to as satellite frame). The satellite frame also can charge to large voltages relative to the environment; hence, large electric fields may exist both at the surface of the satellite and internal to dielectric surface materials. Because of these large electric fields, electrical breakdowns (or discharges) may occur from the charged insulating surfaces to the satellite frame. These discharges can cause electrical interference, which may affect the satellite electronics. It has been suggested that such discharges may be responsible for anomalous behavior observed on satellites that operate at synchronous altitudes.[2,3] In addition, laboratory tests have shown that electrical discharges cause physical damage to the dielectric materials that are used for satellite thermal control.[4] Satellite-charging also can cause the reattraction of ionized contaminants that have outgassed from the interior of the satellite.[5] Physical damage to thermal-control surfaces and the coating of those surfaces with contaminating particles can cause the satellite to operate at an undesirable temperature.

Since these problems were believed to be related to satellite-charging, an experiment built by the Stanford Research Institute (SRI) in Menlo Park, Calif. was flown on a satellite at synchronous altitude to detect these charging events and to detect discharges occurring on the exterior of the satellite as a result of differential charging. This report discusses the in-flight observations made with this experiment and examines the relationship of electrical discharges to the engineering anomalies observed on the satellite.

Instrumentation

This experiment uses a detector that can detect the arrival of energetic electrons at the satellite and the electrical discharges generated by those electrons. The experiment has been described in detail by Nanevicz et al.[6] The following is a brief description of this detector and its function. The detector is a flat metallic plate, 150 cm^2 in area, biased at a fixed voltage of -5.6 v with respect to satellite frame. This detector serves a dual function on the satellite. The total current to the surface of the plate is measured, and electrical pulses (0.5 μsec to 1 msec in length) are counted by the instrumentation. The electronics used to measure the current incident on the detector plate is called the current probe electronics, and the instrument that counts electrical pulses is called the pulse counter.

The current probe electronics measures the total current to the surface of the detector plate. At synchronous altitudes this current is approximately equal to the difference between the photoelectron current from the surface of the detector plate and the plasma-electron current incident on the plate. During quiet geomagnetic conditions, the plasma-electron current is much less than the photoelectron current; therefore, the current measured during undisturbed conditions is essentially equal to the photoelectron current emitted from the detector plate. During substorms, however, the plasma-electron current increases, and the total current measured by the current probe decreases. The electronics measures positive values of current only (electrons leaving the probe) so that quantitative values of the component of plasma-electron current can be measured only as long as it is less than the magnitude of the photoelectron current. For greater values of plasma-electron current, corresponding to a more energetic plasma environment, the instrumentation reads zero current to the probe surface, even though the total probe current may have a negative value. Therefore, the presence of energetic electrons that have been injected into the magnetosphere during substorms can be detected by a decrease (or depression) in the total probe current. Such energetic electrons are believed to be responsible for charging satellites to large negative voltages.

The pulse counter detects voltage pulses on the detector plate caused by electromagnetic radiation from electrical discharges that occur on the surface of the satellite. Laboratory calibrations of this instrument indicate that discharges generated on typical satellite thermal-control surfaces are detected within a range of about 3 ft from the detector plate. The range of detection is not sharp, and discharges with sufficient strength could be detectable anywhere on the satellite's surface. The response of the pulse counter to discharges has been calculated theoretically, and it is in agreement with the laboratory experiments that were performed to determine the range of detection of the instrument.[7] The instrumentation provides no amplitude information; however, it does provide an indication that a discharge has occurred somewhere on the surface of the satellite.

Observations of Electrical Discharges

Figures 1 and 2 illustrate data recorded during both undisturbed and disturbed geomagnetic field conditions for one complete satellite orbit. At synchronous altitudes, each orbit is 24 hr long, and all local times are sampled equally. Data from the current probe are plotted as a function of local time (LT) in the upper panel of Figs. 1 and 2. These data were selected at points in the satellite rotation at which the total probe current was a maximum. (This corresponds to the orientation at which the detector plate is most nearly perpendicular to the sun direction, resulting in the maximum photoelectron current.) The geomagnetic field conditions were determined by examination of A_p and K_p shown at the top of Figs. 1 and 2. (For further information concerning the interpretation of ground-based magnetic indices, see Ref. 8).

Figure 1 shows an example of data recorded during undisturbed geomagnetic field conditions on Jan. 22, 1974. The usual diurnal variation of the peak current detected by the current probe is apparent in Fig. 1. This variation occurs because the photoelectron current is dependent on the angle between the sun direction and the detector plate. At local evening and local morning, the sun direction is nearly perpendicular to the detector plate, and the peak photoelectron current-density has a maximum of about 80 $\mu A/m^2$. At local midnight, the sun direction is at an oblique angle with respect to the detector plate, and the peak photoelectron current-density has a minimum of about 30 $\mu A/m^2$. Near local noon, the body of the spacecraft shadows the detector, and the photoelectron current is reduced nearly to zero, as shown from 10.7 - 13.3 hr LT in Fig. 1. These data are representative of the usual diurnal variation of peak photoelectron current, and no depressions of the probe current indicating the injection of energetic electrons are observed.

Data from the pulse counter are shown in the lower panels of Figs. 1 and 2. The data plotted are the number of dis-

ELECTRICAL DISCHARGE OBSERVATIONS

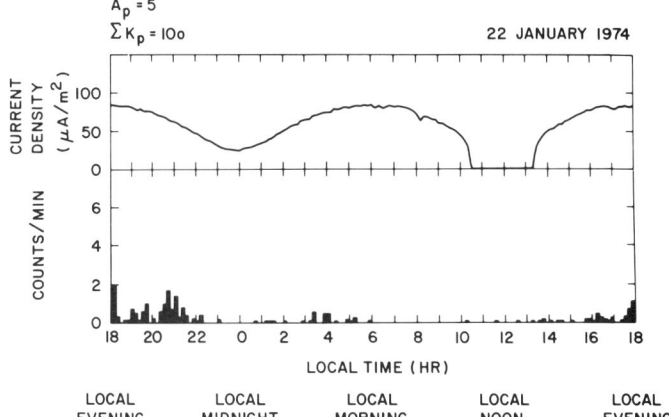

Fig. 1 Static-charge experiment data from an orbit for which no substorm activity was observed. The usual diurnal variation is evident in the current probe data. Few electrical discharges are observed except on the interval from local evening to local midnight.

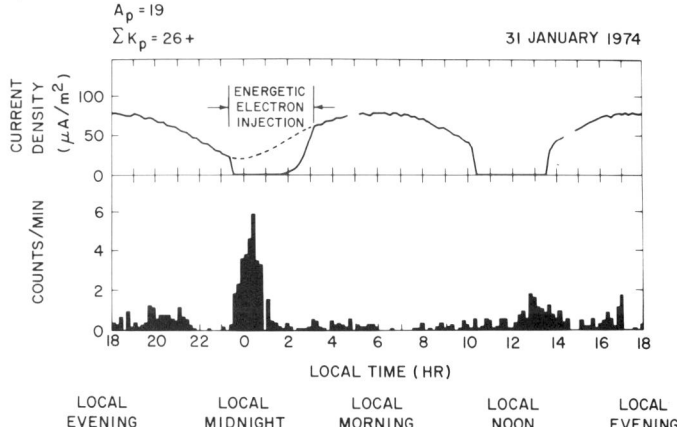

Fig. 2 Static-charge experiment data from an orbit for which substorm activity was observed. Depressions of the probe current near local midnight are caused by the injection of energetic electrons to synchronous altitudes. Discharge rates as high as 6/min are measured by the pulse counter during the electron injection event.

charges per minute that are detected by the pulse counter. The number of counts were collected and averaged over a 10-min period. During the quiet geomagnetic period, shown in Fig. 1, few discharges were detected by the instrument except from local evening to local midnight, during which rates averaging slightly less than 1/min were observed. Count rates, typically near 30/hr, are observed on almost every orbit from local evening to local midnight.

Figure 2 illustrates data recorded during an orbit on Jan. 31, 1974 for which a moderate geomagnetic field disturbance was indicated by the magnetic indices. The injection of energetic electrons at synchronous altitude was detected by observing a depression of the total probe current, which was caused by an increase in the plasma-electron current density during the injection event. Count rates as high as 6/min were observed by the pulse counter near local midnight during the injection event. Discharge rates increased everywhere on this orbit from those measured on the "quiet" orbit shown in Fig. 1, with the exception of the discharges near local evening which occurred at rates similar to those measured on the "quiet" orbit. Discharge rates in excess of 20/min have been observed during magnetospheric substorms; however, a rate of 5-10/min is more typical of events for which depressions of the probe current are observed. Such high rates are observed for extended periods only in the midnight-to-dawn region of the magnetosphere, and only during times when the probe current is depressed, similar to the event shown in Fig. 2.

Two examples of data recorded during satellite eclipse are shown in Figs. 3 and 4. Figure 3 shows the current-probe and pulse-counter data recorded during undisturbed geomagnetic conditions. During satellite eclipse, the probe current from the detector plate drops to nearly zero because there is no photoemission from the detector plate. Smooth changes in the total probe current on either side of satellite eclipse are caused by changes in the satellite orientation (see Fig. 1 and the related discussion of the diurnal variation in total probe current). Only a few pulse counts are observed by the pulse counter during this pass. (The pulse counter data are accumulated and averaged over a 2-min period for the data shown in Figs. 3 and 4.)

Figure 4 shows an example of data recorded during a magnetospheric substorm near satellite eclipse. A depression of the probe current indicating the injection of energetic electrons is observed just after the satellite emerges from eclipse. (Compare these data with those shown in Fig. 3.) Discharge rates as high as 10/min are observed in association with this substorm electron-injection event. The data in Fig. 4 show that the discharges are associated with the injection of the energetic electrons near local midnight, not with satellite eclipse. The beginning of the plasma-injection event shown in Fig. 4 cannot be determined from the current probe data because the satellite is in eclipse. It is possible that

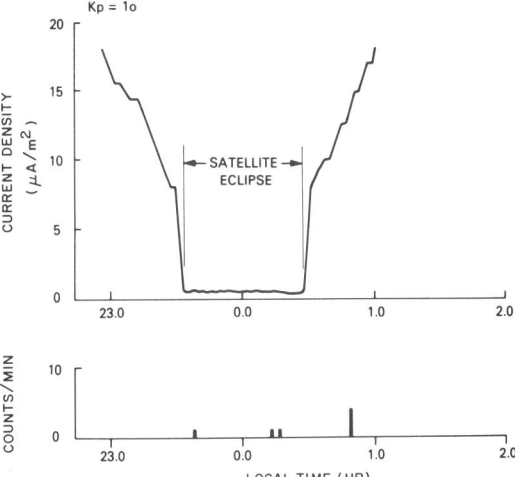

Fig. 3 Static-charge experiment data recorded during undisturbed geomagnetic field conditions through satellite eclipse. The probe current drops to nearly zero when the photoelectric current is cut off by the Earth's shadow. Only a few discharges are observed during this pass.

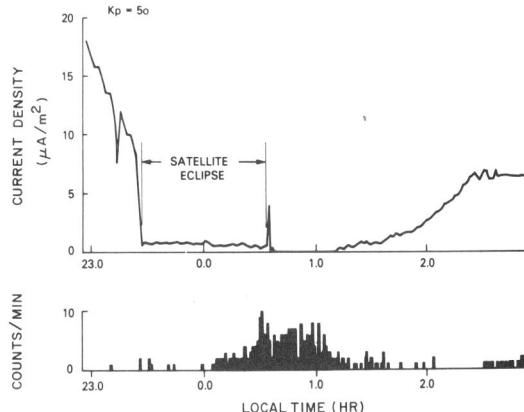

Fig. 4 Static-charge experiment data recorded during disturbed geomagnetic field conditions through satellite eclipse. An energetic electron-injection event and discharge rates as high as 10/min are observed during this pass. The discharges are associated with the injection event (note depressed probe current) rather than with satellite eclipse.

the discharge rates may have been enhanced during satellite eclipse; however, the generating mechanism of the discharges is clearly the energetic plasma injection during the magnetospheric substorm.

These four examples of discharges observed by the static-charge experiment are typical of the several conditions under which synchronous orbiting satellites must function. Electrical discharges are observed during substorms at rates near 10/min on the satellite's external surface. These discharges occur when energetic electrons are injected to synchronous altitudes near local midnight. Such discharges probably are caused by the charging of dielectric surfaces to large negative potentials during the electron injection events, as predicted by others.[2,3] Other discharges are observed near local evening and at other points in orbit at which discharges were not expected to occur. These discharges occur at lower rates (one every 2 min); however, they occur on essentially all satellite orbits from local evening to local midnight and apparently do not require the charging of surfaces to large negative potentials. The generation of these discharges is not well understood at the time of the writing of this paper. It has been suggested that these breakdowns may be related to large electron emission from thin oxide films, which was first observed by Malter.[6,9]

Spin Modulation of Discharges

Figures 5-7 show the distribution of occurrence of these discharges as a function of satellite spin angle. Figure 5 shows the spin modulation of discharges observed during the electron-injection event from 23.5 - 3.0 hr LT on Jan. 31 (see Fig. 2). The angular coordinate in Fig. 5 is the angle of the pulse-counter detector measured in the spin plane of the satellite with respect to the component of the sun direction in that plane. The radial coordinate is the percentage of discharges that occur in a given angular sample interval of about 36°.

The distribution of the occurrence of the local midnight discharges shown in Fig. 5 is nearly isotropic, although there may be a higher probability for detection when the detector is on the shadowed side of the satellite. About 54% of the discharges were observed when the pulse-counter detector was shadowed by the body of the satellite. Since the pulse counter probably cannot detect discharges on the opposite side of the satellite, this may indicate that the discharges occur slightly more frequently on the shadowed side of the satellite.

Figure 6 shows the distribution of the discharges that were observed on Jan. 31 at times for which a depression in the probe current was not observed. Figure 7 shows the distri-

ELECTRICAL DISCHARGE OBSERVATIONS 69

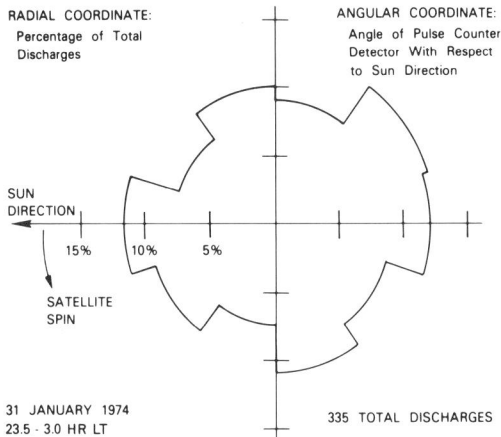

Fig. 5 Spin modulation of substorm produced discharges near local midnight (see Fig. 2).

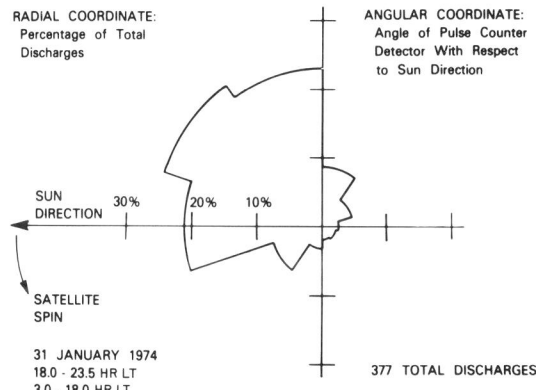

Fig. 6 Spin modulation of discharges observed during a disturbed geomagnetic period at points in orbit for which energetic electrons were not observed (see Fig. 2).

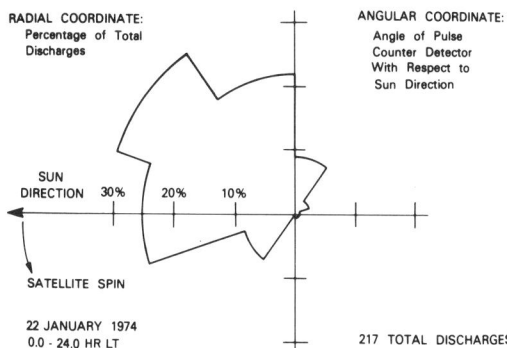

Fig. 7 Spin modulation of discharges observed during a quiet geomagnetic period (Fig. 1).

bution of the discharges observed during a geomagnetically quiet day on Jan. 22, 1974 (Fig. 1). These discharges were recorded during periods for which no depression of the probe current was observed; therefore, for which energetic electrons were not detected by the probe. These distributions are considerably different from that shown in Fig. 5, during which energetic electrons were detected by the current probe. The discharges shown in Figs. 6 and 7 are strongly spin-synchronous. They occur most often at an angle of about 36° before the sun is most perpendicular to the detector surface. This distribution suggests that these discharges are localized on the satellite. They are caused most likely by the periodic shadowing of some particular surface material as the satellite spins.

In summary, the discharges on the satellite appear to be of two distinct types. The first of these occurs during energetic electron-injection events near local midnight. They are nearly isotropic with respect to satellite spin, and they are distributed probably more or less uniformly over large areas of the satellite. The second type are strongly spin-synchronous, and they are found at all points in the satellite orbit. They occur most frequently near local evening, and they occur even during quiet geomagnetic periods.

Variations of Discharge Rate with Geomagnetic Activity

Several examples of orbits on which electrical discharges were observed by the pulse counter have been discussed. The data shown in these examples suggests that these discharges occur at higher rates during periods of increased geomagnetic activity. A survey of data from this experiment was taken during a three week period from Jan. 12 to Feb. 2, 1974. All of the experiment data that was available (74% coverage) were processed for this survey. A study of the relationship of the discharge rates as a function of geomagnetic activity was made from these data. Figure 8 shows the average discharge rate as a function of the geomagnetic index a_p. An equivalent value for a_p was calculated for each 3-hr interval corresponding to the published values of K_p. These values then were averaged over the periods for which the experimental data were available and plotted for each day as a function of the average discharge rate. The resulting index, a_p, corresponds to the daily magnetic index, A_p, when experimental data were available for the entire day. It corresponds to an index similar to A_p averaged only over the period for which experimental data were available on the remaining days.

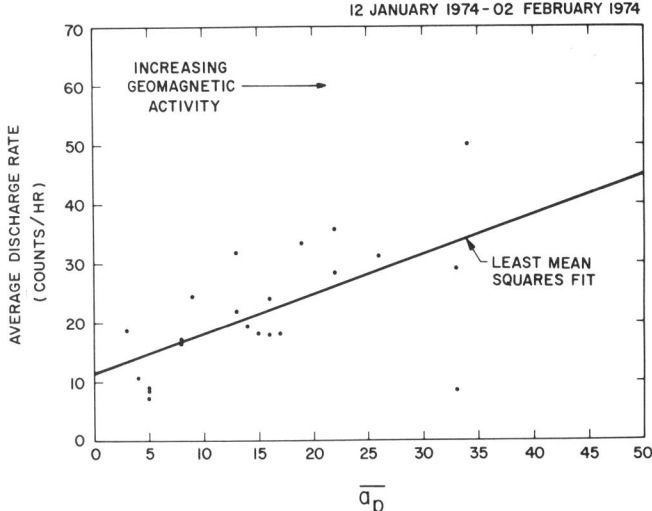

Fig. 8 Average discharge rate as a function of the average value of a_p. These data, averaged over approximately 24 hr, show that the discharge rate increases with increasing geomagnetic activity.

Figure 8 shows that the average discharge rate increases with increasing geomagnetic activity. Figure 8 also shows that during periods of low geomagnetic activity an average count rate of about 10 discharges/hr is observed. Thus, a significant number of discharges are observed during undisturbed geomagnetic periods when essentially no discharges might have been expected. The primary contribution to the count rate during periods of low geomagnetic activity is the discharges that have been observed near local evening on nearly every satellite orbit (see Figs. 1 and 2).

Relationship to Satellite Anomalies

Several types of engineering orbital anomalies have occurred on this satellite which have been supposed to be related to electrical discharges. These anomalies consist of a type of anomalous detector response and an uncommanded reset of a satellite command circuit. The characteristics of these anomalies have been discussed in detail elsewhere and will not be discussed in further detail here.[10,11] These anomalies have occurred on several other satellites in addition to the satellite on which this experiment was flown. An examination of the values of ground-based geomagnetic indices at the times of occurrence of the anomalies strongly suggests that they are related to intense substorm activity.[3] Consequently, it was suggested that such anomalies were generated by interference from the electromagnetic radiation of electrical discharges.

During the period of operation of this experiment, several occurrences of both types of these anomalies were reported. For the anomalous detector response, a total of 35 anomalies were reported. For 31 of these cases, a discharge was observed by the experiment coincidently with the anomalous detector response. These anomalous responses were reported on several different days, and the probability that the coincidences occurred by chance is essentially zero. There were a total of six of the uncommanded circuit resets, and five of these resets occurred coincidently with detection of discharges by the experiment. Therefore, data from this experiment have confirmed the hypothesis that these types of orbital engineering anomalies are caused by interference from these electrical discharges.

These discharges may also have a damaging effect on the passive thermal control of the satellite. Discharges generated in the laboratory with high-energy electron beams have caused damage to typical satellite thermal control surfaces. Such damage could result in degradation of the satellite thermal performance if it occurs on-orbit. The temperature rise rate of a structural element on this satellite was plotted as a function of the average value of D_{ST} to determine if a relationship existed between magnetospheric substorm activity and the satellite warming trend. Figure 9 shows that a correlation does exist between the temperature rise rate and the geomagnetic index D_{ST} averaged over the interval for which the rise

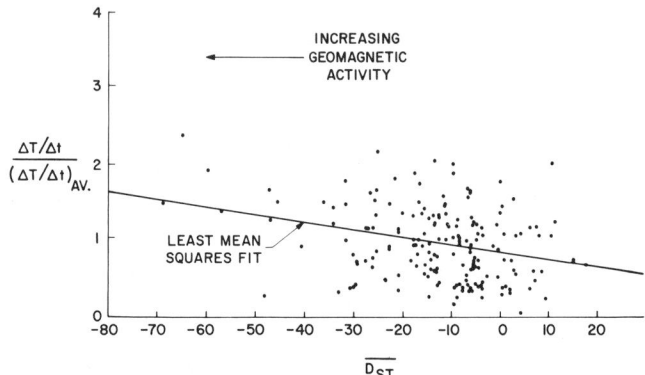

Fig. 9 A correlation exists between increases in the temperature of sensors on these satellites and increases in magnetospheric substorm activity as determined by D_{ST}. This suggests the possibility that part of this temperature rise could be caused by phenomena associated with satellite charging.

rate was measured (several days). Although the data points show considerable scatter, there is a clear trend for higher rise rates to occur at most disturbed (more negative) values of D_{ST}. A detailed statistical analysis of these data showed that there was less than 0.1% probability that there is actually no relationship between the temperature rise rate and D_{ST}.

The exact mechanism causing this correlation is unknown, and the total amount of degradation that can be attributed to mechanisms other than substorms is unknown. Physical damage of the type observed on thermal-control materials in laboratory charging experiments may explain the correlation shown in Fig. 9. It is impossible, however, to estimate realistically the amount of degradation caused on-orbit without further knowledge of the particular materials and satellite structures actually affected in-flight.

While the statistical analysis indicates that there is almost certainly a relationship between the temperature rise rate and D_{ST}, the correlation between them is weak (linear correlation coefficient equal to approximately 0.25). This weak statistical correlation does not necessarily mean that discharges cause only a small amount of degradation to the thermal performance of the satellite for the following reasons. First, the discharges occur at daily average rates near 10/hr even when magnetospheric conditions would be expected to be quiet (see Fig. 8). The daily average discharge rate increases only by a factor of 3 or 4 during disturbed magnetospheric conditions even though large increases in the discharge rate are observed over a small portion of the orbit near local midnight during energetic plasma injection events. Therefore, it is possible that a significant amount of the thermal degradation could be caused by the discharges and that no unusually dramatic increases in the temperature rise rate would be observed during periods of strong geomagnetic activity. Second, it must be emphasized that the geomagnetic indices, such as D_{ST}, K_p, and A_p, are inherently statistical in nature. They are computed from ground-based magnetogram records and do not necessarily indicate environmental conditions at the satellite that would result in increased discharge rates. For example, if the satellite is not near local midnight during an energetic plasma injection event, the discharge rate would not be expected to rise as much as if the satellite had been at the proper orbital position to encounter these energetic electrons.

In order to assess the importance of magnetospheric substorms as a cause of the degradation in the satellite passive thermal control, it is necessary to study the effect of the environment on a particular satellite at synchronous altitudes. The correlation between the temperature rise rate of this satellite and D_{ST} suggests that satellite charging could play a significant role in degrading the thermal performance of dielectric materials on satellites at synchronous altitudes.

Conclusions

The data obtained from the static-charge experiment has verified that electrical breakdowns (or discharges) do occur on the surface of a satellite at synchronous altitude. In addition, these discharges appear to be of two types, categorized by two distinguishable sets of characteristics. The first type of discharge occurs in the presence of energetic electron injection events near local midnight. These discharges occur more-or-less uniformly on both the sunlit and shadowed sides of the satellite at rates of about 10/min over the area for which a discharge can be detected by the instrumentation. These discharges are probably caused by large negative charging of the spacecraft when it encounters energetic plasma injected to synchronous altitude during magnetospheric substorms. The second type of discharge is strongly synchronized to the spin of the satellite. They are observed at all orbital locations, with the highest rates (1/min) near local evening, and they are observed during both quiet and disturbed geomagnetic conditions. These discharges may be caused by periodic shadowing of some particular material on the surface of the satellite, or they may be related to the Malter emission observed on thin oxide films.[6,9]

Two types of engineering orbital anomalies on these satellites were observed to occur coincidently with discharges observed by the pulse counter. This observation confirms the hypothesis that these engineering anomalies are caused by electromagnetic interference from electrical discharges. In addition, a statistical comparison of the anomalous temperature rise rate of a satellite structural element with the geomagnetic index D_{ST} suggests that there is a relationship between the degradation of the satellite thermal performance and substorm activity. This relationship may occur because of physical damage to the dielectric surfaces caused by discharges. A determination of the contribution of such damage to the degradation of a satellite's passive thermal control will require further study on satellites at synchronous altitude.

These problems, all of which are related to satellite-charging, point out the necessity for understanding the mechanism of satellite-charging, the discharge process, and the effects of electrical discharges on satellites. Much work remains to be done before the generation of such discharges can be understood and controlled. Some analysis of the coupling of discharge electric fields to satellite electronic circuits with suggested techniques for reducing electrical interference from the discharges has been performed.[12,13] Experimental in-flight data from future satellites will be necessary to determine the importance of these discharges as a mechanism for degrading satellite thermal control surfaces.

References

[1] DeForest, S. E., "Spacecraft Charging at Synchronous Orbit," Journal of Geophysical Research, Vol. 77, Feb. 1972, pp. 651-659.

[2] Fredricks, R. W. and Scarf, F. L., "Observations of Spacecraft Charging Effects in Energetic Plasma Regions," Photon and Particle Interactions with Surfaces in Space, D. Reidel Publishing Co., Dordrecht, Holland, 1972, pp. 277-308.

[3] Cauffman, D. P., "Correlation of Strobe Anomalies with Geophysical Parameters," The Aerospace Corp., El Segundo, Calif., Rept. ATM-74(4409-04)-3, 1973.

[4] "Mirror-Panel Static Charge Tests," Aerojet ElectroSystems Co., Azusa, Calif., Rept. 4656, 1972.

[5] Cauffman, D. P., "Ionization and Attraction of Neutral Molecules to a Charged Spacecraft," The Aerospace Corp., El Segundo, Calif., TR-0074(9260-09)-1, 1973.

[6] Nanevicz, J. E., Adamo, R. C., and Scharfman, W. E., "Satellite Lifetime Monitoring," Stanford Research Institute, Menlo Park, Calif., SRI Project 2611, 1974.

[7] Cauffman, D. P. and Shaw, R. R., "Transient Currents Generated by Electrical Discharges," Space Science Instrumentation, Vol. 1, Feb. 1975, pp. 125-137.

[8] Handbook of Correlative Data, J. H. King, Editor, The National Space Science Data Center, NASA/Goddard Space Flight Center, Greenbelt, Md., 1971, pp. 76.

[9] Malter, L., "Thin Film Field Emission," Physical Review, Vol. 50, July-Dec. 1936, pp. 48-58.

[10]"Orbital Data Analysis Report," Vol. II, Aerojet Electro-Systems Co., Azusa, Calif., Rept. 4362-10, April 1973.

[11]"Orbital Data Analysis Report," Aerojet ElectroSystems Co., Azusa, Calif., Rept. 4362-11, June 1973.

[12]"Geomagnetic Substorm Effects Study," TRW Systems Group, Redondo Beach, Calif., Rept. 16439-90-049-601, Oct. 1972.

[13]Shaw, R. R., "Geomagnetic Substorm Activity Study," Aerojet ElectroSystems Co., Azusa, Calif., Rept. 5120, Feb. 1975.

SUBSTORM-INDUCED SPACECRAFT-CHARGING CURRENTS FROM FIELD-ALIGNED AND OMNIDIRECTIONAL PARTICLES

J. L. Vogl,* N. L. Sanders,*
TRW Defense & Space Systems, Redondo Beach, Calif.

and

S. E. DeForest+
University of Alabama, Huntsville, Alabama

Abstract

Recent observations by ATS-6 of intense field-aligned fluxes of electrons at geosynchronous orbit have been compared with omnidirectional fluxes seen at the same time. These calculations show that the field-aligned component cannot be ignored in spacecraft-charging calculations. Differential charging of insulation at the bottom of narrow cavities might be dominated entirely by the field-aligned fluxes when the spacecraft is in an appropriate orientation. Such considerations might explain why anomalies induced by spacecraft charging are not associated with geomagnetic disturbances on a one-to-one basis.

Introduction

Anomalous behavior induced by spacecraft charging has been shown to affect most classes of spacecraft flown at geosynchronous orbit.[1,2] Examples of such anomalous behavior include command changes, spurious noise in sensors, and degra-

---

Presented as Paper SA36 at the American Geophysical Union National Meeting, Washington, D. C., June 1975. This work was supported in part under Air Force Contract No. FO 4701-69-C-0091. The authors wish to thank A. Rosen, G. Inouye, and C. McIlwain for helpful discussions.
  *Member of the Professional Staff.
  +Associate Professor of Physics.

dation of solar cells, optical surfaces, and thermal-control surfaces. In a few cases, subsystem failures also have been linked to charging. Many difficulties so far have prevented a complete understanding of these phenomena. The first is that the environment is not very well known. No spacecraft has been flown at geosynchronous orbit (GSO) with a full complement of particle and field instruments. And, even if the environment were well specified, simulation in the laboratory would be very difficult. The second major difficulty is that most spacecraft that suffer anomalies at GSO carry no environmental monitoring instruments. Therefore, correlation must be done with ground-based observations and with other spacecraft. This combination of really not understanding the environment and not having sufficient instrumentation leads to the third difficulty: that anomalies do not occur on a one-to-one basis with geomagnetic activity. Some spacecraft seem to be able to operate perfectly for months. They suffer no effects due to the many minor substorms or occasional large storms that they encounter. Then, for no readily apparent reason, they start getting false commands during geomagnetic activity that is no stronger than normal.

This lack of one-to-one correspondence is so startling that some investigators have rejected spacecraft charging as the problem in spite of other circumstantial evidence, such as the local times of occurrence of anomalies which are concentrated in the midnight-to-dawn region (where plasma temperatures and fluxes are particularly high[3]) and the observation that it is necessary, but not sufficient, to have had geomagnetic activity before the anomaly.

In this paper, we explore the possibility that intense field-aligned fluxes of electrons recently reported by McIlwain[4] could be a new parameter that must be specified in order to describe the environment accurately enough to predict anomalous behavior. To do this, we estimate the currents to a spacecraft from the omnidirectional component and compare this to the field-aligned component.

Measurements

1) Omnidirectional

The bulk of the particles responsible for electrostatic charging of spacecraft at GSO have been measured by two University of California at San Diego (UCSD) instruments on ATS-5 and ATS-6. An earlier instrument[5] on ATS-1 failed before an eclipsing period was encountered and large potentials seen. The ATS-5 instrument detects electrons and ions in the energy range of 50 eV to 50 keV in two directions, parallel and perpendicular to the spin axis of the spacecraft. As a result of this geometry, plus the location of ATS-5 at 105°W, neither detector normally looked to within 20° of the loss cone.

Only some very early data of limited duration were taken when the spacecraft was tumbling end over end. In spite of the fact that the data taking was averaged over approximately 100°, definite field-aligned fluxes of particles were seen.[6] However, the data were compromised severely. Some of the ATS-5 particle experimenters used the tumble period to extract pitch angle information.[7,8] However, this was not possible with the UCSD ATS-5 instrument. As a result, the UCSD plasma instrument for ATS-6 was redesigned to sweep mechanically in angle to hunt for field-aligned fluxes.

Since the ATS-5 data have been analyzed for several years and are reasonably well understood, we shall use them as a data base for omnidirectional fluxes. The normal method of presentation for studying magnetospheric dynamics is the spectrogram (see Ref. 3 for examples) but for this study we desire only certain moments, and not the whole distribution function. Therefore, we have calculated the effective temperature and particle flux (i.e., current) for a representative active and quiet period. The results are shown in Figs. 1 and 2. Note that, because of changing spectra, the peaks in current and temperature do not come necessarily at the same time. In both of these figures, the temperature has been calculated by taking an average energy over the measured spectrum. However, previous studies have shown that the energy distribution of the newly injected particles has a nearly Maxwellian shape. This is shown in Fig. 3, which is taken from the previously cited paper by DeForest and McIlwain.[3] Knowledge of this temperature is necessary to predict induced potentials, since the spacecraft will adjust its potential until the net flux to it is identically zero. Figure 4 shows how the total current available from the environment is decreased if only particles of certain energies are permitted to reach the surface. Neglecting effects such as shocks, wakes, and the plasma sheath about the spacecraft, and assuming that one knows the composition of the external surfaces, one then can use the following equation with Figs. 1, 2, and 4 to predict the spacecraft potential:

$$J_{total} = J_p - J_e + J_{BS} + J_{Se} + J_{Sp} + J_{pe} \qquad (1)$$

where J_p is the current density, J_e is the electron current density, J_{BS} is the backscattered electron current density,

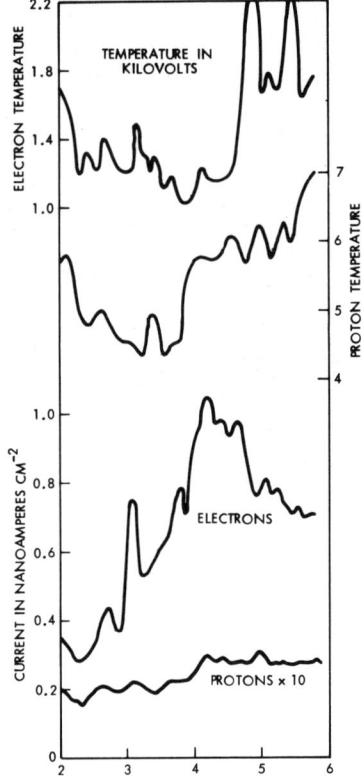

Fig. 1 Temperature and current profiles for March 27, 1970 substorm.

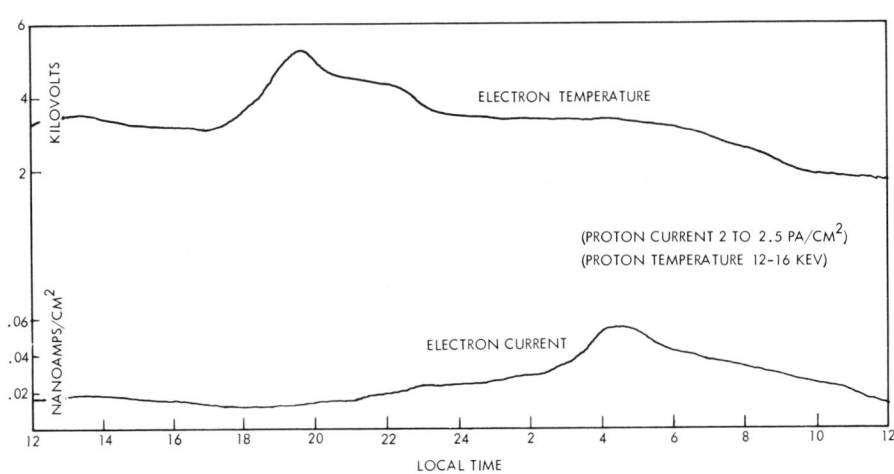

Fig. 2 Electron temperature and current profiles during typical quiet time. (February 6-7, 1970 on ATS-5.)

FIELD-ALIGNED AND OMNIDIRECTIONAL CHARGING CURRENTS   81

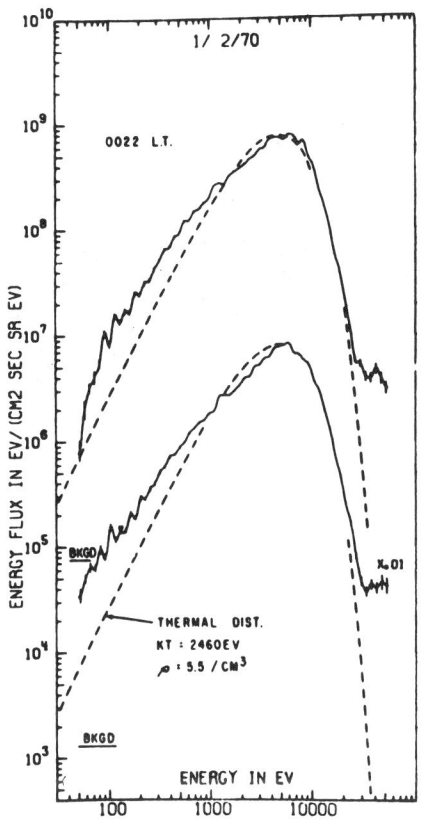

Fig. 3 Maxwell-Boltzmann fit to electron distribution data from ATS-5.

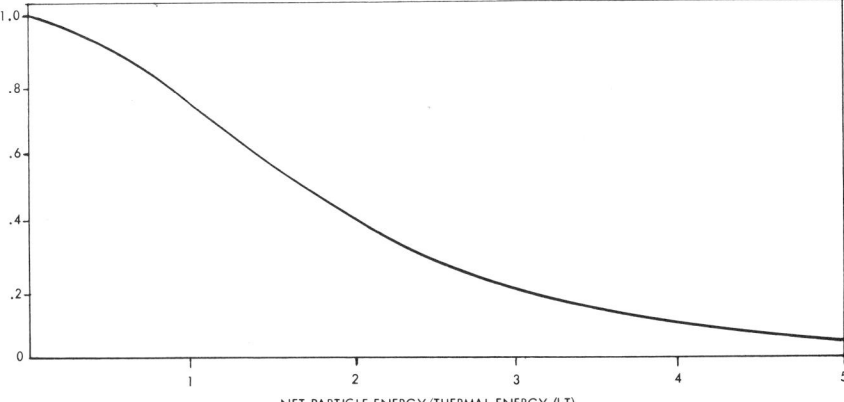

Fig. 4 Fraction of particles reaching a surface as a function of net particle energy for a Maxwell-Boltzmann distribution.

J_{S_e} is the secondary electron current density from electrons, J_{S_p} is the secondary electron current density from protons, and J_{pe} is the photoelectron current density. This approach has been used by DeForest[9] to predict ATS-5 potentials. A more advanced model, which attempts to predict differential charging with the use of the same data base, has been presented by Inouye.[10] No detailed analysis or model development will be done here. But, having presented the omnidirectional component likely to be seen at GSO, we now discuss the field-aligned component to see if Eq. (1) must be modified to include these extra terms.

2) Field-Aligned

The UCSD plasma instrument onboard ATS-6 covers a broader energy range (0.1 eV to 80 keV) and has provisions to sweep out angles in the north-south plane and in the east-west plane by rotating the sensing heads mechanically. Even with this flexibility, the north-south head rarely sweeps through the loss cone. This happens because, at the initial location of ATS-6 at 94° W long, the magnetic field is inclined to the swept north-south plane by 10-20°. After ATS-6 moves to the eastern station at 35° E long, then the magnetic field will lie in the swept plane. For this reason, statistics on the occurrence of field-aligned fluxes that have been observed to date must be viewed as lower limits only. As McIlwain points out,[4] ATS-5 was unable to detect field-aligned fluxes in over 4 yr of normal operation because it was not looking in the right direction. ATS-6 does better, but, since the classical loss cone is only 3.5° at GSO, and since the orientation of the magnetic field varies both with geomagnetic activity and spacecraft location (i.e., local time), it is difficult to locate without an onboard real-time feedback from the magnetometer.

Approximately 20 field-aligned events were seen in the first six months of operation. These lasted from less than a minute to about 20 min in duration. The characteristic energy varied from hundreds to thousands of electron volts. Two representative cases are shown in Figs. 5 and 6. Since ATS-6 was at 94° W at these times, Fig. 5 was taken at approximately 1700 LT and Fig. 6 at approximately 0000 LT. Note that the central energies are quite different in the two cases.

These figures were constructed by allowing the instrument to sweep separately through or near the loss cone while simultaneously stepping asynchronously in energy. In this way, a series of tracks through pitch angle - energy space were built up. Contours then were passed through these tracks to connect points of constant counting rate. Approximately 10 to 20 min

FIELD-ALIGNED AND OMNIDIRECTIONAL CHARGING CURRENTS 83

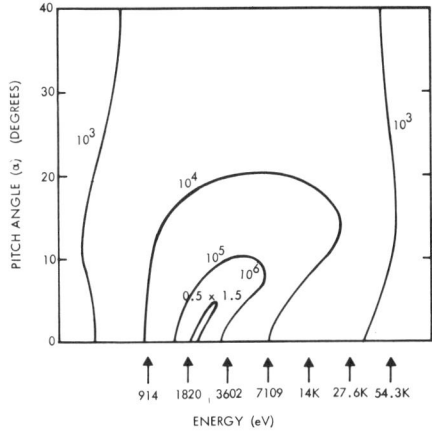

Fig. 5 Electron isocountrate contours in pitch angle energy plane ATS-6, day 233 (0625-0646 UT).

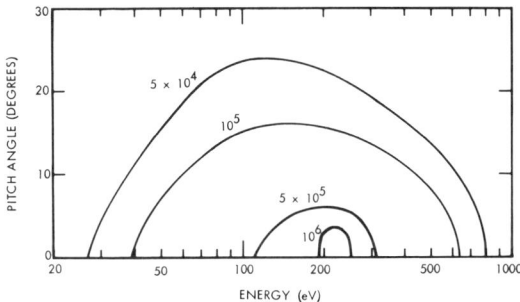

Fig. 6 Electron isocountrate contours in pitch angle-energy plane ATS-6, day 186 (2254-2304 UT).

is needed to collect sufficient data for this process. Therefore, one assumption in these figures is that the spectra are time-stationary for periods of that order. We know from McIlwain's work cited earlier[4] that there are factors of 2 fluctuations in the field-aligned component with time scales of less than 1/4 sec. Furthermore, a detailed study of the raw data in Fig. 6 indicates that it might really consist of two separate events. Therefore, these curves should be used only to get a general picture of the shape of the field-aligned component and to estimate the total flux. Although the geophysical cause of the events is a very interesting topic of current research, for the purposes of this paper we shall simply accept the data in Figs. 5 and 6 and not speculate on the processes that produce them.

Even with these limitations, some information on the shape of the energy and pitch angle distributions can be made which simplify the computation of the electron current densities.

The pitch angle distributions at peak energy, obtained from the isocountrate contours, are shown in Fig. 7. In estimating the contribution of the field aligned electrons to the spacecraft charging current it was assumed that the pitch angle distribution was a decaying exponential. Note the close approximation of an exponential fit to the data as shown in Fig. 7.

The differential energy spectra at zero pitch angle also was obtained from the contour plots. These are shown in Figs. 8 and 9. The flux in each case has a peak, over a relatively narrow energy range. Except for this peak, the energy spectra can be approximated roughly by a Maxwellian energy distribution. This suggests that for computational purposes, the actual energy distribution of field aligned electrons can be replaced by two energy components; a monoenergetic electron group and a component having a Maxwellian energy distribution. Note that, in making Figs. 7-9 from Figs. 5 and 6, a geometric factor of 1.6×10^{-4} cm^2-sterad was used for the electron channels. These geometric factors are defined with the energy

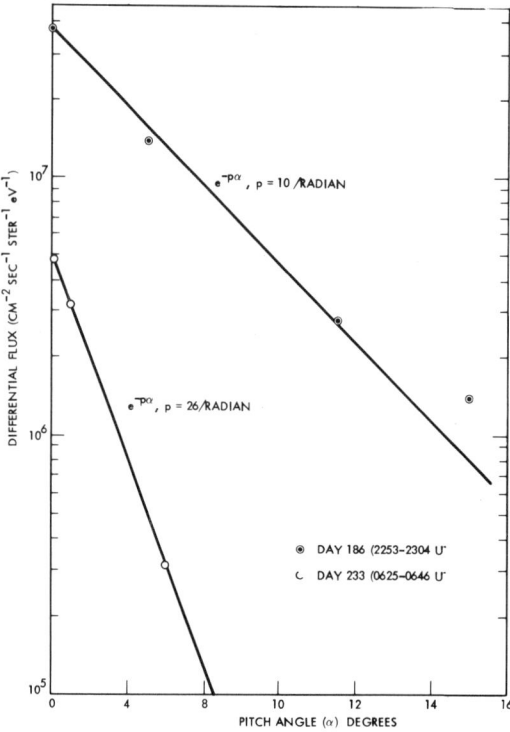

Fig. 7 Pitch angle distributions at energy of peak flux.

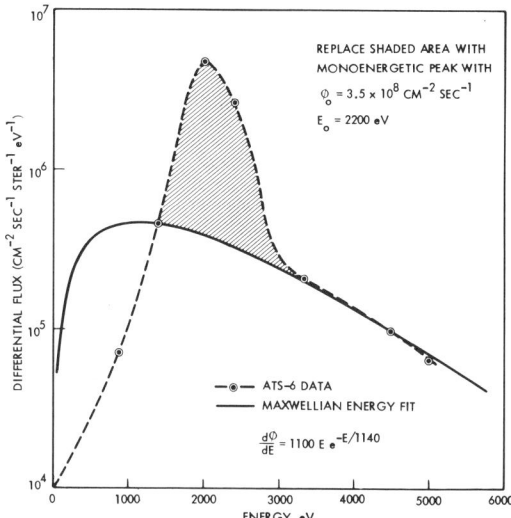

Fig. 8 ATS-6 differential energy spectra at zero pitch angle, day 233 (0625-0646 UT).

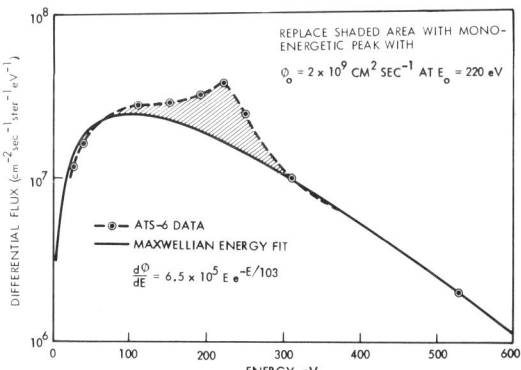

Fig. 9 ATS-6 differential energy spectra at zero pitch angle, day 186 (2255-2304 UT).

bandwidth folded in so that the unidirectional flux is obtained from the countrate in the following manner:

$$j(\text{part/cm}^2\text{-sterad-sec-eV}) = C(\text{sec}^{-1})/E(\text{eV})H(\text{cm}^2\text{-sec}),$$

where H is the geometrical factor and E, the energy.

Results and Discussion

Using the functional fit developed in the previous section for the field-aligned flux on days 186 and 233, along with the

Table 1 Currents to ATS-6 from field aligned and omnidirectional particles

	Day 186	Day 233
Omnidirectional current density	14 na/cm^2	5.0 na/cm^2
Field-Aligned current density	0.32 na/cm^2	0.06 na/cm^2
Cavity L/D ratio for equal currents	1.6	2.3
Cavity L/D ratio for field-aligned to exceed omnicurrent by a factor of 10	5.2	7.4

measured omnidirectional flux and temperature as shown in the section before last, the total currents to ATS-6 were calculated assuming no large spacecraft potentials (which was the case for both events). The results are shown in Table 1. We can see immediately the field-aligned electron contribution to the total spacecraft current is very small. This confirms the approximations used in previous models, which assume only omnidirectional fluxes.

Even though the field-aligned contribution to the total current is relatively small, if a spacecraft cavity is oriented in the direction of the field during such an event, the charging effects may be significant. For example, we have calculated the ratio of depth to diameter for a cylindircal cavity such that the omnidirectional and field-aligned components illuminate the bottom in equal amounts. For isolated cavities deeper than this amount, we would expect the field-aligned fluxes to dominate the charging equations when the cavity is positioned with its axis along the local magnetic field line. These ratios also are shown in Table 1. Also shown are the ratios at which the field-aligned flux is 10 times the omnidirectional fluxes. Both calculations are based on geometry only and do not include scattering paths or internally produced secondaries.

Many spacecraft configurations have cavities with these diameter-to-depth ratios. The deeper they are, the more unlikely it is that they ever will be aligned with the magnetic field when there is an intense field-aligned flux. However, if the bottom of such a deep cavity contains insulators or sensitive components, its response will be determined entirely by the field-aligned component. As we have seen from Fig. 6, this could mean that the bottom of a deep cavity could charge to thousands of volts, whereas the main spacecraft frame resides at a much lower potential determined almost entirely by the omnidirectional flux and photoemission. This process then could explain why some spacecraft survive very strong geometric disturbances only to suffer anomalous behavior during smaller events at different times. This also might explain the weak seasonal occurrence of some types of anomalies in addition to the diurnal effect.

This analysis is valid only for spacecraft with characteristic dimensions of the order of 10 m or less. Reasoner et al.[11] discuss the possibility that the nature of spacecraft charging effects might change at geosynchronous orbit when the dimensions increase to the order of 10 km. There is a possibility that these field-aligned fluxes could create an instability in the plasma near such a large vehicle. This could be relevant to recent proposals to build large solar-power spacecraft.[12,13] Because of the great uncertainty in the occurrences and strength of the intense field-aligned fluxes, much of what we suggest here borders on speculation. However, we feel that at least we have shown that the recently reported intense field-aligned fluxes of electrons at geosynchronous orbit can be a hazard to spacecraft. Furthermore, we feel that the existence of these fluxes helps to explain the lack of one-to-one correspondence between geomagnetic activity and anomalous behavior.

References

[1] McPherson, D. A., Cauffman, D. P., and Schober, W., "Spacecraft Charging at High Altitudes - the SCATHA Satellite Program," AIAA Paper 75-92, Jan. 20-22, 1975, Pasadena, Calif.; published elsewhere in this volume.

[2] Rosen, A., "Spacecraft Charging—Environment-Induced Anomalies," AIAA Paper 75-91, Jan. 20-22, Pasadena, Calif.

[3] DeForest, S. E. and McIlwain, C. E., "Plasma Clouds in the Magnetosphere," Journal of Geophysical Research, Vol. 76, No. 16, June 1971, pp. 3587-3611.

[4] McIlwain, C. E., "Auroral Electron Beams near the Magnetic Equator," Nobel Symposium, April 1975, Kiruna, Sweden. (To be published by Plenum Press, London.)

[5] Freeman, J. W., Jr. and Maquire, J. J., "Gross Local-Time Particle Asymmetries at the Synchronous Orbit Altitude,"

Journal of Geophysical Research, Vol. 72, No. 21, Nov. 1967, pp. 5257-5264.

[6]DeForest, S. E. and McIlwain, C. E., "Anisotropic Pitch Angle Distribution of 50 eV to 50 keV Particles at Synchronous Altitude," COSPAR, 1972, Madrid, Spain.

[7]Sharp, R. D., Shelley, E. G., and Johnson, R. G., "Preliminary Results of a Low-Energy Particle Survey at Synchronous Altitude," *Journal of Geophysical Research*, Vol. 75, No. 31, Nov. 1970, pp. 6092-6101.

[8]Bogott, F. H. and Mozer, F. S., "Equatorial Proton and Electron Angular Distribution in the Loss Cone and at Large Angles," *Journal of Geophysical Research*, Vol. 76, No. 28, Oct. 1971, pp. 6790-6805.

[9]DeForest, S. E., "Spacecraft Charging at Synchronous Orbit," *Journal of Geophysical Research*, Vol. 77, No. 4, Feb. 1972, pp. 651-659.

[10]Inouye, G. T., "Spacecraft Charging Model," AIAA Paper 75-255, January 20-22, 1975, Pasadena, Calif.; also, *Journal of Spacecraft and Rockets*, Vol. 12, Oct. 1975, pp. 613-620.

[11]Reasoner, D. L., Lennartsson, Walter, and Chappell, C. R., "Relationship between ATS-6 Spacecraft-Charging Occurrences and Warm Plasma Encounters," presented as Paper SA-36A at the American Geophysical Union National Meeting, Washington, D.C., June 1975; published elsewhere in this volume.

[12]Woodcock, G. R. and Gregory, D. L., "Derivation of a Total Satellite Energy System," AIAA Paper 75-640, April 24, 1975, Los Angeles, Calif.

[13]Glaser, P. E., Maynard, O. E., Markoviciak, J., Jr., and Ralph, E. L., "Feasibility Study of a Satellite Solar Power Station," NASA Rept. CR-2357, Feb. 1974.

RELATIONSHIP BETWEEN ATS-6 SPACECRAFT-CHARGING OCCURRENCES AND WARM PLASMA ENCOUNTERS

D. L. Reasoner,[*] Walter Lennartsson,[†] and C. R. Chappell[≠]

NASA Marshall Space Flight Center, Huntsville, Ala.

Abstract

The occurrences in local time of spacecraft-charging and warm plasma events observed by the University of California at San Diego plasma detector on ATS-6 have been examined using data from 40 contiguous orbits. The local time distribution of charging events was found to maximize between local midnight and dawn and was also similar to the distribution of synchronous spacecraft anomalies reported by others. More than half of the charging events had potentials of -50 V or more, and potentials on the order of -1000 V were observed. All of the spacecraft charging events reported in this paper occurred during sunlit conditions. The warm plasma encounters were concentrated in the local noon-to-dusk sector, and the local-time distributions of charging events and warm plasma encounters were found to be anticorrelated.

Introduction

The study of spacecraft-charging is motivated both for scientific and practical reasons. Scientifically, it is impor-

---

Presented as Paper SA-36A at the American Geophysical Union National Meeting, Washington, D.C., June 1975. We are indebted to C. E. McIlwain of the University of California at San Diego for making the ATS-6 auroral particles experiment data available to us. This work was supported, in part, by NASA RTOP 385-36-01 and was done while two of the authors (D. L. R. and W. L.) were NAS/NRC postdoctoral fellows at NASA Marshall Space Flight Center.
[*]NAS/NRC Senior Postdoctoral Fellow.
[†]NAS/NRC Postdoctoral Fellow.
[≠]Chief, Magnetospheric and Plasma Physics Branch.

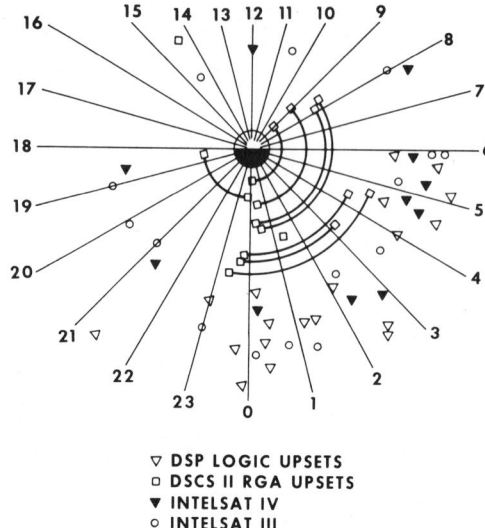

Fig. 1 Distribution in local time of anomalies observed on various geosynchronous satellites.

tant to understand interactions between plasmas and surfaces in space. As a practical matter, understanding the spacecraft environment and its effects upon spacecraft operations is no less important. One example of such a practical motivation is displayed in Fig. 1, from McPherson et al.[1] This figure shows the distribution in local time of various types of "unexplained" anomalies observed on spacecraft at synchronous orbit. It is seen that, rather than being distributed randomly, there is a strong tendency for the anomalies to cluster in the local time period of midnight to dawn. Considering that several spacecraft are represented in this plot and that some peculiar effect due to sun angle probably would not persist for all of the satellites, it seems reasonable to conclude that, in fact, there is some geophysical phenomenon that is concentrated in this local time period which leads directly or indirectly to spacecraft anomalies.

Evidence that the culprit is spacecraft-charging was provided by studies of charged particle fluxes on the geosynchronously orbiting ATS-5 satellite by DeForest and McIlwain[2] and by DeForest.[3] The manifestation of a magnetospheric substorm at synchronous orbit was shown to be a sudden injection of clouds of hot plasma, and these plasma cloud particles subsequently would drift according to their energy and charge sign, producing characteristic dispersion curves in energy-time space. DeForest[3] found that there was a definite association between ATS-5 spacecraft-charging and the occurrence of substorm-injected hot plasmas, and that the satellite surface potentials were particularly high when the satellite was in eclipse. Given, then, that potentials on the order of a few thousand volts can occur on spacecraft at synchronous orbit, it was only reasonable to conclude that dielectric breakdown of various satellite surface materials could occur. These breakdowns would cause significant electromagnetic interference that could have upsetting or disastrous effects upon satellite electronic circuitry.

The basic physical law governing the potential of a free body in a plasma environment is simply that the net current to the surface must be zero. An unbalanced current would result in a charge imbalance which would in turn create a surface potential of sufficient magnitude to restore current balance. For a body in a plasma, there are several sources of current, and the current balance equation can be written as

$$J_e + J_i + J_{pe} + J_{sec} + J_m = 0$$

where J_e and J_i are the currents due to external plasma electrons and ions, J_{pe} is the net photoelectron current, J_{sec} is the net current due to secondary electrons generated by energetic primaries at the satellite surface, and J_m is a possible contribution from man-made sources on the satellite, such as ion and electron guns, plasma arcs, or thermionic emitters. Parenthetically, these last sources offer a means of controlling the satellite potential. In any case, all of these current sources are a function of the satellite surface potential and the distribution of potential between the satellite surface and the undisturbed plasma at large distances. Therefore, for a given photon and plasma environment, the satellite potential adjusts to the value required to satisfy the current balance equation. Consider, for example, the case of a free body in a vacuum illuminated by the sun. In this case, all current terms except J_{pe} are zero, and, hence, the net photoelectron current must be identically zero. Therefore, a photoelectron sheath forms around the body, and conservation of energy requires that the surface potential be equal to the energy of the highest-energy photoelectron generated. This effect has been observed in the case of the moon by Reasoner and Burke[4] where the surface potential in the low-density regions of the Earth's geomagnetic tail was measured as 200 V positive with a lunar-surface-based plasma detection instrument. Another simple case involves a free probe in a plasma composed of electrons and ions of the same temperature, and no photoelectron sources present. Since the electron thermal velocity is $(m_i/m_e)^{1/2}$ times the ion thermal velocity, there is a much higher probability for an electron to strike the surface than for an ion. Therefore, the probe must take on a negative potential in order to reduce the electron current so that it will be equal to the ion current.

In a "zeroth-order" treatment of the problem of the potential of a satellite in typical space conditions, the dominant terms may be considered to be the external plasma electron current and the photoelectron current. If the available photoelectron current exceeds the external plasma electron current, the satellite potential will be relatively small and positive. If the converse is true, then the satellite will charge negatively to a value required to reduce the external plasma electron current to a value required for current balance. Since the substorm-injected plasmas at synchronous orbit have electron temperatures on the order of thousands of volts, we now can understand how the ATS-5 satellite could charge to such large negative potentials, particularly during eclipses when the photoelectrons were absent. Yet, as will be seen, the remaining terms in the current balance equation can make significant contributions, particularly when $J_e \approx J_{pe}$ and therefore can have a large effect upon the potential. To obtain a more complete understanding of the effects of the plasma environment upon spacecraft-charging, it is necessary to make measurements of the complete plasma environment in the vicinity of the spacecraft.

The ATS-6 spacecraft offered an opportunity to make further studies of the phenomenon of spacecraft-charging and its relationship with the external plasma environment. The satellite carried an environmental measurements experiment (EME), consisting of several plasma detectors and a magnetometer. In this paper, we present observations from the University of California at San Diego auroral particles experiment of the occurrences of spacecraft-charging and its relation to the plasma environment.

Description of the Experiment

The ATS-6 satellite was launched into geosynchronous orbit on May 28, 1974 and was maintained on station at 94° west long until May 15, 1975. The orbital inclination is <2°. The configuration of the satellite is sketched in Fig. 2. The satellite is oriented so that the parabolic antenna points toward the Earth, and the solar cell panel support booms lie perpendicular to the orbital plane (approximately north-south). The EME package is visible on the back side of the antenna.

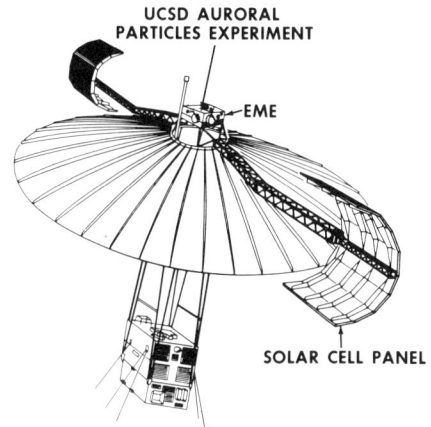

Fig. 2 Sketch of the ATS-6 satellite.

A complete description of the UCSD auroral particles experiment will be forthcoming in a future publication. A brief description will suffice for the purposes of this paper. The instrument consists of two mechanically rotating head assemblies, each of which contains a pair of electrostatic deflection sensors. One sensor is sensitive to ions, and the other is sensitive to electrons. In addition, there is a fixed sensor, sensitive to ions only, which looks in an easterly direction. The two rotating heads rotate in orthogonal north-south and east-west planes, and hence the nomenclature for the five data channels is 1) north-south ions; 2) north-south electrons; 3) east-west ions; 4) east west electrons; and 5) fixed east ions. The fields of view of the sensor is 5° x 5°, and the angular sweeping range is 220°.

The energy scanning of the sensors is controlled by a common voltage reference generator. The analyzed energy per unit charge alternates between a scan through 64 levels between 0 and 81,000 eV in 16 sec, and dwell steps at a fixed energy. An overall scan-dwell program is set up by ground command, and there is complete flexibility in choosing the dwell steps and the length in time of the dwell steps. Additional commands are available to control the angular sweeping mode of the rotating heads and the accumulator gating. In the so-called normal accumulator gating mode, which is the most common mode, each of the four rotating sensors has a 0.25-sec accumulation time, and the fixed east sensor has a 0.125-sec accumulation time.

Data

Motivated by the scientific and practical reasons to study spacecraft-charging, and the improved sensitivity to low-energy plasma of the ATS-6 instrument relative to the ATS-5 instrument, a study of the occurrence of ATS-6 spacecraft-charging and its relation to other phenomena was undertaken. Forty contiguous days from June 15 to July 24, 1974 were selected for study, based on data availability at the time the study was undertaken. This data base, including more than one solar rotation period, covered a wide range in magnetic and substorm activity ranging from extremely quiet to very disturbed days. Admittedly, however, the data base was too short in duration to discern any long-term seasonal effects.

The primary thrust of the investigation was to examine the data for the occurrence of spacecraft-charging and its relation to substorm activity, and for the presence of high-density regions of low-energy or "warm" plasma, which would be the result of encounters of ATS-6 with the plasmasphere. Previous studies of the plasmasphere with Ogo-5 (see Chappell[5] and references therein) were with an instrument that measured only the density of the plasma. The ATS-6 instrument afforded the opportunity to measure both the density and temperature of the low-energy plasmaspheric plasma.

The next series of figures will show examples of the various phenomena referred to earlier in this section. The basic ATS-6 data display format with which this study was done was a series of 1-min line plots. Each 1-min frame consisted of data from the five sensors, data from the magnetometer, which was kindly supplied by R. L. McPherron of UCLA, and various ancillary data including the energy scan, dwell steps, and the look angles of the rotating heads. For purposes of presentation, we have extracted from the line plots only the ion and electron data from the north-south head and the scan-dwell step data. Figure 3 shows an example of these plots. The top panel shows the counting rate due to ions, the middle panel the counting rate due to electrons, and the bottom panel is the energy-step indicator. At this particular time, ATS-6 was at 18:37 local time (LT = UT - 6 hr), and the only significant external particle fluxes are a quiet-time ion ring current with a density of ≈ 0.2 ion/cm^3 and a temperature of ≈ 30 keV. Low-energy photoelectrons from the spacecraft itself are seen also. This last is seen as the narrow peak in the electron data at low energies.

Figure 4 shows an example of spacecraft-charging. Here the spacecraft is immersed in a hot electron plasma due to an injection event that occurred approximately 1 hr earlier. The spike in the ion data at E = 120 eV indicates the presence of a negative spacecraft charge of 120 V. The spike is the result

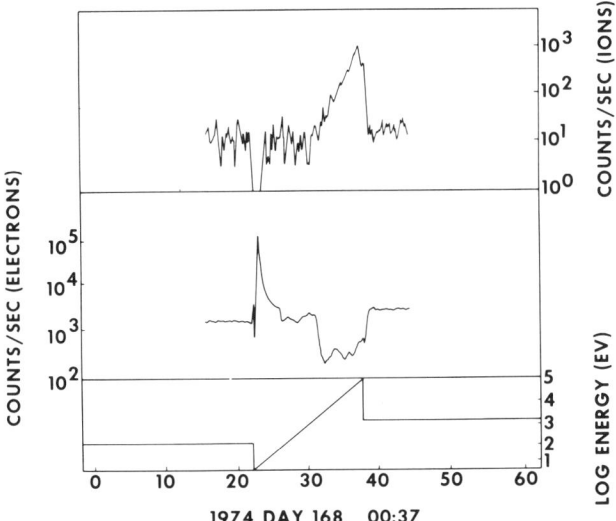

Fig. 3 Example of ATS-6 plasma data for a period of no charging. The data traces are explained in the text.

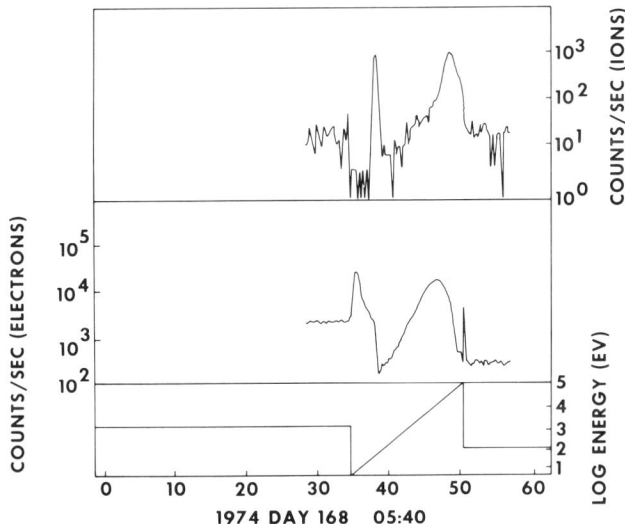

Fig. 4 Example of spacecraft-charging indicated by the spike in the ion data. Here the spacecraft potential is -120 V.

of lower-energy ion plasma, which has been accelerated by the negative spacecraft potential. We also see that, for energies between 0 eV and the spacecraft potential, there are essentially no ion counts as is demanded by conservation of energy of ions originally at large distances with initial velocity vectors toward the satellite.

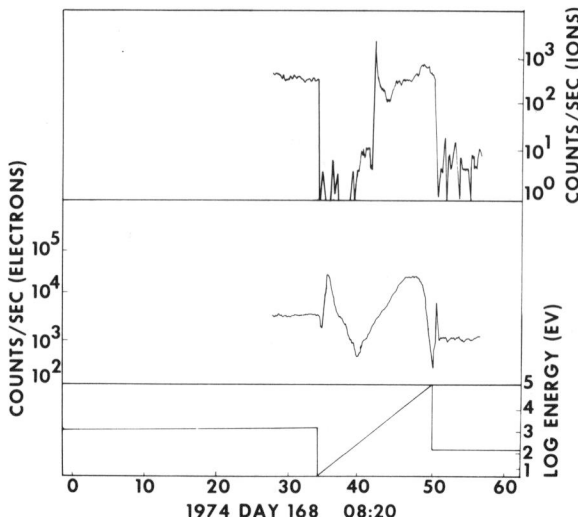

Fig. 5 Example of spacecraft-charging as in Fig. 4. The spacecraft potential is -1400 V.

An extreme example of charging is shown in Fig. 5. Here the same comments apply as in Fig. 4, except that now the spacecraft potential is 1400 V negative. It should be emphasized that, in these examples and for all of the data in this paper, the spacecraft was in sunlight. This shows that even in sunlight when photoelectrons are available to provide return current, the spacecraft can charge to potentials on the order of 1000 V. Thus, high potentials are not only an eclipse phenomenon. Positive satellite potentials of magnitudes less than about 30 V would be impossible to detect by the technique described above because of the contamination of the ambient electron fluxes by photoelectrons. However, no evidence for positive potentials greater than 30 V has been seen in the data.

The preceding two figures also illustrate a feature of the sheath that forms around the charged spacecraft. We see that, even when the satellite is charged negatively, there are still significant fluxes of photoelectrons returning to the spacecraft. This shows that the potential distribution between the spacecraft surface and large distances is in fact nonmonotonic, and that an undershoot potential exists to trap photoelectrons in a layer around the spacecraft. We do not know if this undershoot potential is a common feature of spacecraft-surface interactions or whether it is unique to ATS-6. The primary ramification of the formation of the undershoot potential is, of course, that part of the photoelectron population is trapped near the spacecraft and does not contribute return

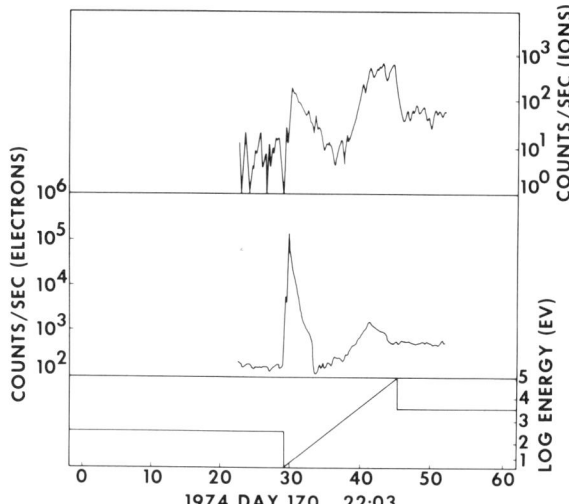

Fig. 6 Example of an encounter with low-energy plasmaspheric plasma, as seen by the enhancement in the ion detector counting rates at low energies.

current to the ambient plasma. This would lead to higher negative potentials than expected from the simple flux balance theory. Whipple[6] has recently discussed the theory of satellite sheaths and its application to the observed ATS-6 potential distributions.

Figure 6 shows an example of a plasmasphere encounter. The low-energy ion component has a density of 2 ions/cm^3 and a temperature of 4 eV. The electron component of this plasma also can be seen in the electron sensor data, but there is strong contamination by local photoelectrons.

To study the systematics of the occurrences of spacecraft-charging and its relation to other phenomena, the 40 24-hr data sets were divided into 10-min segments, and the probability of each type of occurrence was computed as the ratio of the number of occurrences in each 10-min local time interval to the total number of observational periods. The data were plotted in the form of circular histograms, with the angle being proportional to local time and the radius being proportional to probability of occurrence. The plot of the occurrences of spacecraft-charging to negative potentials is shown in Fig. 7. The data set represents all charging events for which magnitude of the negative potential was greater than 10 V, which was the lower limit that could be detected reliably. The occurrence probability is seen to maximize in the local midnight-to-dawn sector, and the similarity of this plot with the plot of the

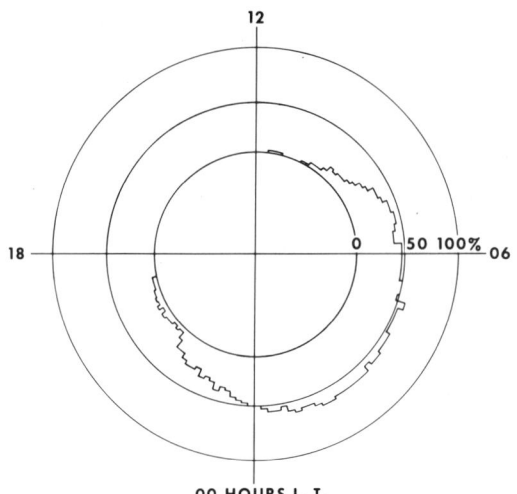

Fig. 7 Local time distribution of ATS-6 spacecraft charging events.

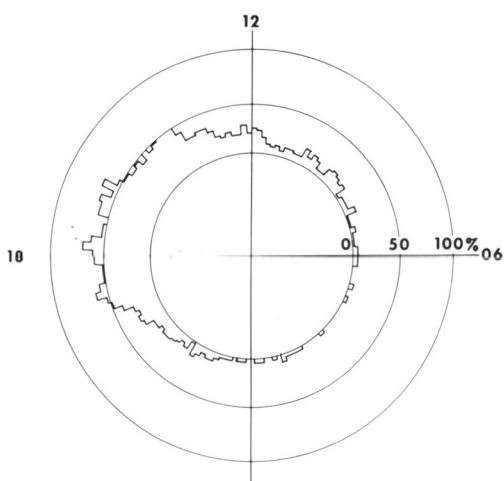

Fig. 8 Local time distribution of ATS-6 encounters with low-energy plasmaspheric plasma. For these events, ion densities ranged between 1 and 10 ions/cm^3.

local time distribution of spacecraft anomalies shown in Fig. 1 is indeed striking. This figure also reinforces the view that charging is the direct result of spacecraft encounters with substorm-injected hot electron plasmas, for these plasmas are injected near local midnight and subsequently are transported by particle drifts toward local dawn. For a more complete discussion of the behavior of substorm plasmas, the reader is referred to Ref. 2.

The distribution of plasmasphere encounters is shown in Fig. 8. These encounters are concentrated in the local noon to local evening sector, and in fact the set of charging events and the set of plasmasphere encounters are anticorrelated on the basis of local time of occurrence. The explanation for this is twofold. First, the high-density, low-energy plasma is an effective source of current to the spacecraft and hence can prevent the potential from reaching large values. Secondly, comparison of plasmasphere encounter data with indices of geomagnetic activity shows that plasmasphere encounters are much more likely during geomagnetically quiet periods. Substorm activity and the consequent hot plasma injections are associated with geomagnetically disturbed periods.

The distribution of the magnitude of the potential was studied by dividing the set of all charging events into 50 V increments. The histogram of percentage occurence vs potential is shown in Fig. 9. Note that the percent occurrence scale is logarithmic. Over half of the charging occurrences were potentials less than 50 V in magnitude, but it is obvious that there is a significant probability for occurrence of potentials >1000 V in magnitude, and the potential reached 2200 V in one case.

Conclusions

The statistical study of ATS-6 spacecraft-charging reported here reinforces the conclusions of DeForest,[3] McPherson

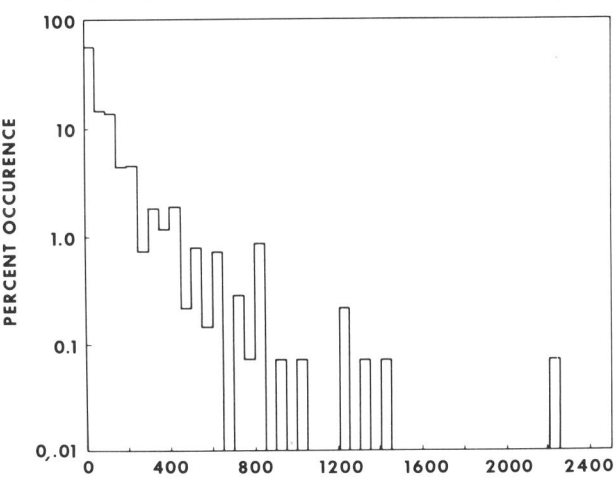

Fig. 9 Distribution of ATS-6 spacecraft potentials in charging events.

et al.,[1] Rosen,[7] and others that spacecraft-charging at synchronous orbit is a direct result of the injection of hot electron plasmas during magnetospheric substorms. The probability of spacecraft-charging was found to maximize in the local midnight to local dawn sector. The low-energy measurement capability of the UCSD instrument allowed a survey of the relation between charging and encounters with the low-energy plasmaspheric plasma. Spacecraft-charging and encounters with the plasmasphere were seen to be mutually exclusive. Another factor affecting charging was seen to be the formation of an undershoot potential that served to prevent a large fraction of the photoelectrons from leaving the spacecraft and thus contributing to the return current.

ATS-5 charging observations[3] show a maximum negative potential in sunlight of ≈ -200 V. By contrast, the potential of ATS-6 was -200 V or greater for approximately 10% of the observed charging events, and in one extreme case reached -2200 V. This illustrates that the presence of photoelectrons does not, a priori, guarantee that the spacecraft potential will be clamped to relatively small values.

It should be kept in mind also that the ATS-6 potentials reported here are, in fact, the potentials of the immediate vicinity of the EME package relative to the plasma at large distances. There is no information about differential charging effects. Such effects undoubtedly must occur. The satellite is constructed of many different materials, both conductors and insulators, and the materials are exposed to varying amounts of solar radiation in the course of the orbit. Furthermore, as Vogl et al.[8] show, the electron distribution function occasionally is peaked strongly along magnetic field lines. This would result in parts of the spacecraft receiving larger fluxes of energetic electrons than others.

Recently it has come to the authors' attention that feasibility studies are underway in various organizations concerning power generation and transmission satellites in geosynchronous orbit with scale sizes on the order of kilometers. The possible problems due to the interaction of such large spacecraft with the plasma environment are interesting to contemplate. Any such system probably would involve large collector areas with one surface in sunlight and the other in darkness. In addition, this scale size is smaller than a typical ion gyroradius but larger than a typical electron gyroradius. This would lead to geometrical particle shadowing; that is, all parts of the surface would not be exposed to identical environmental plasma fluxes. There would be differentation according to the charged particle energy, charge sign, and direction. Consequently, this would contribute to differential charging.

The experience with ATS-5 and ATS-6 shows that the magnitudes of spacecraft-charging potentials are not only a function of the local plasma environment but also are influenced strongly by the details of the satellite construction. The ability to predict the behavior of a satellite suffers from the absence of a complete theory of spacecraft-charging, sheath formation, and of the requisite data on photoemissivity and secondary electron yield of typical satellite materials. Therefore, one should exercise caution in attempting to predict the charging behavior of a satellite from the limited data presently available.

References

[1] McPherson, D. A., Cauffman, D. P., and Schober, W., "Spacecraft Charging at High Altitudes - The Scatha Satellite Program," AIAA Paper 75-92, Jan. 1975, Pasadena, Calif.

[2] DeForest, S. E. and McIlwain, C. E., "Plasma Clouds in the Magnetosphere," Journal of Geophysical Research, Vol. 76, June 1971, pp. 3587-3611.

[3] DeForest, S. E., "Spacecraft Charging at Synchronous Orbit," Journal of Geophysical Research, Vol. 77, Feb. 1972, pp. 651-659.

[4] Reasoner, D. L. and Burke, W. J., "Characteristics of the Lunar Photoelectron Layer in the Geomagnetic Tail," Journal of Geophysical Research, Vol. 77, Dec. 1972, pp. 6671-6687.

[5] Chappell, C. R., "Recent Satellite Measurements of the Morphology and Dynamics of the Plasmasphere," Reviews of Geophysics and Space Physics, Vol. 10, Nov. 1972, pp. 951-979.

[6] Whipple, E. C., "Theory of the Spherically Symmetric Photoelectron Sheath: A Thick Sheath Approximation and Comparison with the ATS-6 Observation of a Potential Barrier," Journal of Geophysical Research, Vol. 81, Feb. 1976, pp. 601-607.

[7] Rosen, A., "Large Discharges and Arcs on Spacecraft," Astronautics and Aeronautics, Vol. 13, June 1975, pp. 36-44.

[8] Vogl, J. L., Sanders, N. L. and DeForest, S. E., "Substorm-Induced Spacecraft-Charging Currents from Field-Aligned and Omnidirectional Particles," published elsewhere in this volume.

SPACECRAFT POTENTIALS IN A
SUBSTORM ENVIRONMENT

G. T. Inouye*

TRW Defense & Space Systems, Redondo Beach, Calif.

Abstract

The potentials of different parts of a spacecraft due to charging from the environmental plasma are calculated to show that the potential differences, which are generated across various capacitive configurations of the spacecraft, may represent significant multikilovolt stresses across energy storage elements (capacitors). Voltage breakdown in or around these capacitors has been postulated as the source of electromagnetic interference (EMI), which causes anomalous behavior of synchronous orbit satellites. Stresses in different parts of the spacecraft are computed by modeling the entire spacecraft with an equivalent electrical network to show seasonal dependences which are consistent with observed anomalies of several different types. The model responses to both step-function and real-time substorm data from ATS-5 show charging times in the order of tens of minutes and discharge times of many hours. The results are strongly dependent on spacecraft configuration, solar direction and materials parameters.

Introduction and Background

The occurrence of serious and potentially serious abnormal behavior of operational communications spacecraft in geosyn-

---

Presented at the 56th Annual AGU Meeting, Washington, D. C., June 1975. Performed under Air Force Contract No. FO 4701-69-C-0091. The author acknowledges the support of J. L. Vogl and N. L. Sanders of the Space Sciences Dept. under A. Rosen in the formulation of the plasma charging currents as well as in many other aspects of the work reported here. The support of J. A. Durschinger of the 777 Project Office, and the painstaking care of Mrs. B. Benefield in the many revisions of the manuscript are greatly appreciated.
*Member of the Professional Staff

chronous orbit has been studied since 1971.[1-6] In-orbit measurements on ATS-5 by S. E. DeForest[7,8] in the same time frame showed that the energetic plasma in the ambient environment could cause spacecraft charging to multikilovolt levels. In an earlier paper,[9] the author developed a charging model for the DSCS II spacecraft with which voltage stresses were obtained as a function of location. These locations, correlated with the positions of electronic hardware on the spacecraft, related to specific types of anomalous behavior. Furthermore, the seasonal dependence of the stress magnitude at each location was demonstrated to be consistent with the actual occurrences of each type of anomaly. Both the locations of high stresses and their seasonal dependence were shown to be influenced strongly by the particular spacecraft configuration and its orientation relative to solar illumination. Another result of the analysis was to show that minor modifications of the spacecraft could reduce or eliminate stresses at several critical locations. These modifications were incorporated into the follow-on spacecraft of the same generic configuration. This paper represents one aspect of the continuing work being performed to understand the phenomenon and to improve the immunity of spacecraft to environmental effects. The spacecraft charging model is reviewed briefly, and a summary of the results obtained previously are presented as an introduction to the material presented here.

Spacecraft Configuration

Figures 1 and 2 are photographs of the DSCS II spacecraft which depict their general configuration. A notable and important feature is that less than 2% of the total external surface area is metallic and tied to the common spacecraft ground. Structural pipes, which constituted the major portion of this metallic area, are shown in the underside view of Fig. 2. The solar array panels are removed in this photograph. Nearly all exterior surfaces are covered or wrapped with thermal blankets, second surface mirrors, solar cell cover glasses, paint, etc. Subsequent examination of a number of other spacecraft configurations has shown that the small proportion of exposed metallic surfaces relative to surface areas is a common feature that has been dictated primarily by thermal control considerations.

Figure 3 shows the seasonal and diurnal variation of the (projected) metallic area exposed to sunlight. The seasonal variation mainly is caused by the structural pipes, seen in Fig. 2, which are sunlit only during the winter months. The diurnal variation mainly is due to the daily rotation of the antennas resulting in the exposure of the waveguides in front of the two large dishes. Note that the percentage variations in areas exposed to sunlight are large, but that these areas

Fig. 1   DSCS II spacecraft.

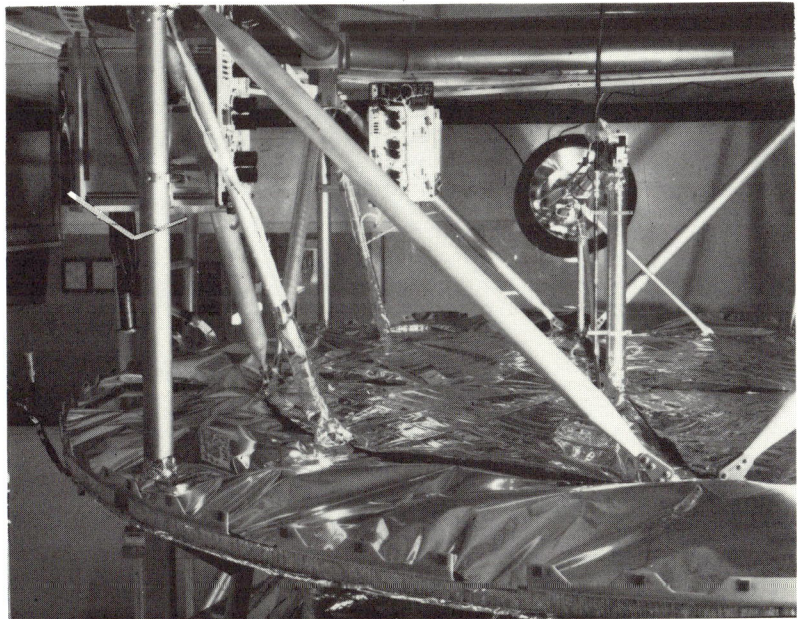

Fig. 2   Underside view of the DSCS II with solar panels removed.

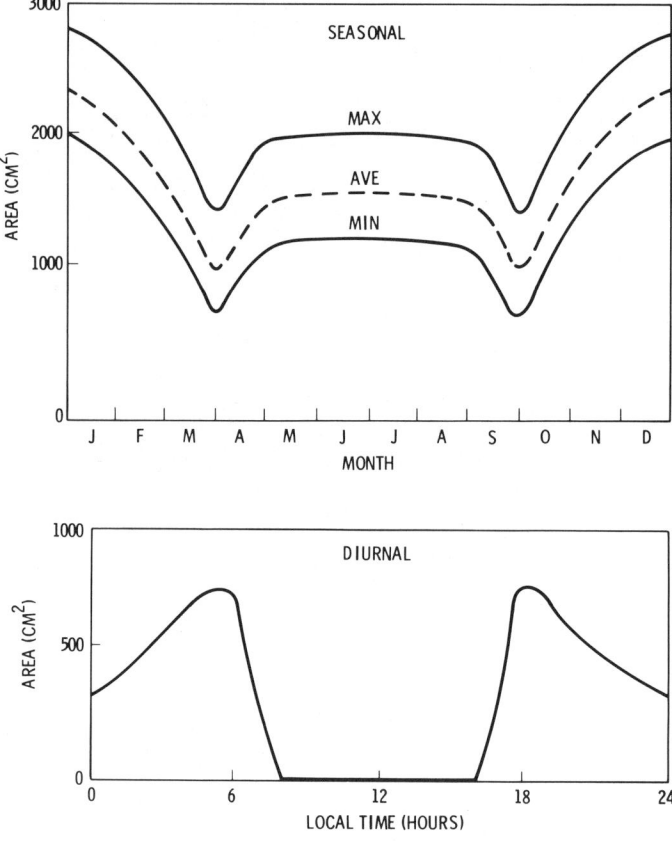

Fig. 3 Seasonal and diurnal variation of metallic areas exposed to sunlight.

are only about 0.7%, on the average, of the 24-m² total exterior surface area. The metallic area exposed to the omnidirectional plasma is about three times larger and does not have a time dependence.

Figure 4 is a schematic representation of the spacecraft for purposes of showing charging potentials at various locations as a function of the seasonal changes of the sun aspect angle. Figure 4a applies to the original configuration (Mod 0) and Fig. 4b to the modified configuration (Mod 2) in which the metallic pipes of Fig. 2 have been covered with thermal blankets and all apertures to the interior volume, in which most of the electronic hardware are located, have been baffled. The steady-state potential differences $(V_n - V_o)$ are compared in

SPACECRAFT POTENTIALS IN A SUBSTORM ENVIRONMENT

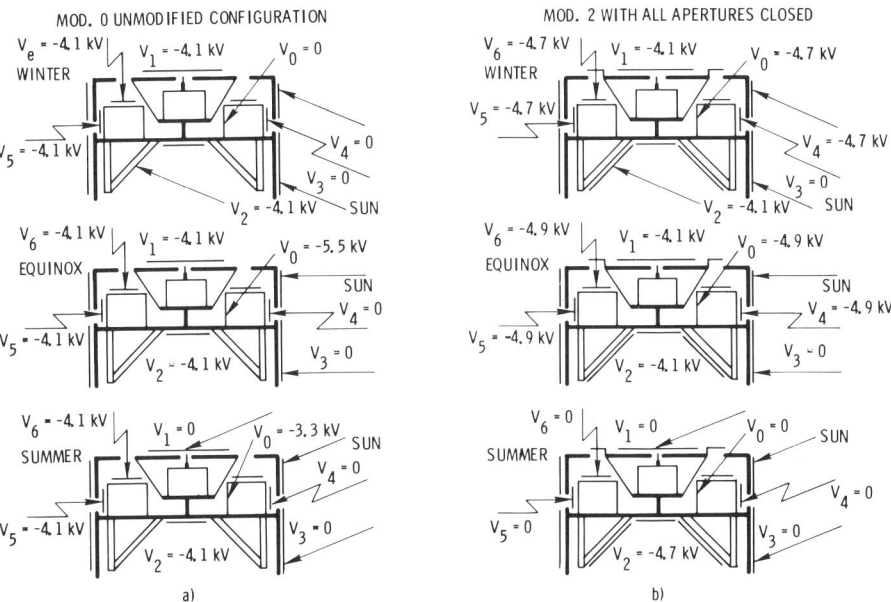

Fig. 4 Schematic diagram of the DSCS II spacecraft showing local potentials as a function of season: a) Mod 0 unmodified configuration, and b) Mod 2 with all apertures closed.

Table 1 for the two configurations. The main features to be noted in Table 1 are that the (V_1-V_0) stress, which has been identified as being critical, has been reduced greatly at all seasons, as have all of the aperture related stresses.

Spacecraft Charging Model

The charging model shown in Fig. 5 is an electrical-circuit representation of the DSCS II spacecraft configuration. The basic metallic structure and all of the metallic objects tied to it constitute the reference mode for the circuit. Its potential, spacecraft "ground", is designated as V_0. Various dielectric and ungrounded metallic surfaces exposed to the ambient plasma are represented as "leaky" capacitors. Each dielectric surface is shown as a separate capacitor. One end of each capacitor is tied to the reference (V_0) node. The other end of each capacitor, as well as the V_0 node itself, constitute nodes to which current generators are connected representing the charging and discharging sources, i.e., plasma electrons and protons, photoemission, and secondary emission. The return node for all of these current generators is the zero potential reference at infinity.

		WINTER			EQUINOX			SUMMER		
	MOD	V_1-V_0	V_2-V_0	V_3-V_0	V_1-V_0	V_2-V_0	V_3-V_0	V_1-V_0	V_2-V_0	V_3-V_0
a	0	-4.1	-4.1	0	1.4	1.4	-5.5	3.3	-0.8	3.3
	1	-2.4	-2.4	1.7	-1.4	-1.4	2.7	0	-4.1	0
	2	0.6	0.6	4.7	0.8	0.8	4.9	0	-4.1	0

		APERTURES CLOSED	WINTER			EQUINOX			SUMMER		
	MOD.		V_4-V_0	V_5-V_0	V_6-V_0	V_4-V_0	V_5-V_0	V_6-V_0	V_4-V_0	V_5-V_0	V_6-V_0
b	0	NONE	0	-4.1	-4.1	5.5	1.4	1.4	3.3	-0.8	-0.8
		6	4.7	0.6	0	4.9	0.8	0	0	-4.1	0
	M2	4,6	4.6	0	0.6	0	0	0.8	0	0	-4.1	0
		4,5,6	0	0	0	0	0	0	0	0	0

MOD. 0 = Original configuration
MOD. 1 = Pipes wrapped only (5462 cm^2)
MOD. 2 = Pipes wrapped (5462 cm^2), metal exposed (3406 cm^2)
SURFACE 0 = Metalic structure (spacecraft ground)
SURFACE 1 = Forward closure
SURFACE 2 = Bottom of spinning platform
SURFACE 3 = Solar cell cover glass
APERTURE 4 = Portholes and slit on solar panels for sunlight
APERTURE 5 = Portholes and slit on solar panels for plasma
APERTURE 6 = Annular gap on forward closure for plasma
V_n = Potential of surface n where n = 0, 1, 2, 3, 4, 5, 6

Table 1 Potential differences, kv, at a) external surfaces and b) apertures

In general, the magnitudes of the circuit components values, capacitance, and leakage conductance are proportional to the areas involved. Figure 5 lists the typical per unit area parameters assumed for each material. Figure 6 shows a preliminary set of experimental data points obtained for 2 mil Kapton thermal blanket material, which indicates that the effective bulk conductivity of Kapton, 10^{+18} (Ω-cm^{-1}), exhibits a threshold electrostatic stress of about 240 kv/cm.[10] Above this threshold, the conductivity increases as the 3.5th power of stress (current as the 4.5th power of stress).

In an examination of the particular spacecraft discussed here, 18 different capacitors (dielectric surface areas) were identified. Only four of these were used in the circuit analysis, since they represented the majority of the total area and would control the reference node potential V_0. Potentials of surfaces not included in the basic analysis were calculated with the assumption that V_0 is not affected by these smaller

SPACECRAFT POTENTIALS IN A SUBSTORM ENVIRONMENT

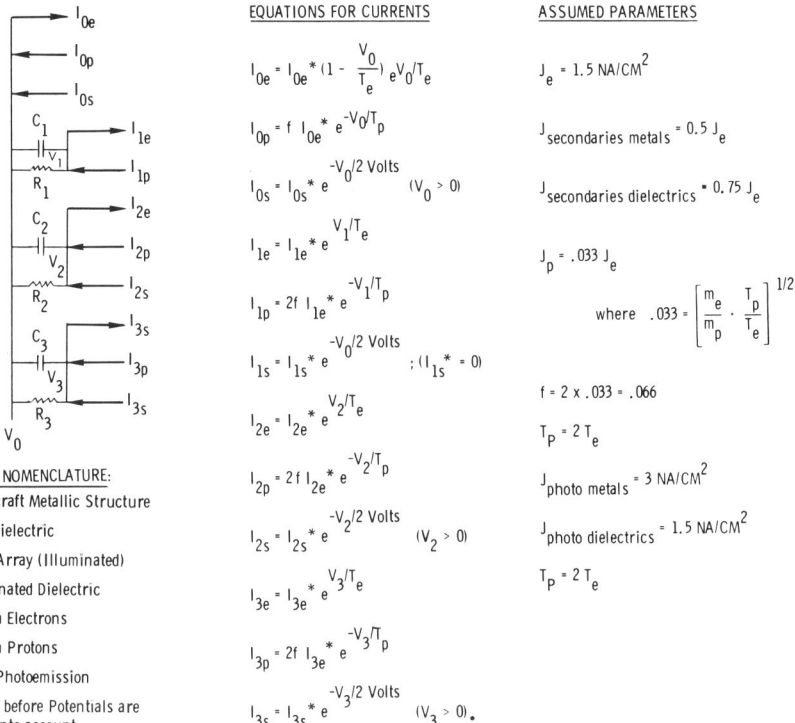

Fig. 5 Equivalent circuit with functional relations.

contributors. Aperture/cavity potentials, for instance, were calculated in this way.

Only three capacitors are shown in the equivalent circuit of Fig. 5 because two of them were combined to form C_1 which represents all dielectric surfaces that are not illuminated by solar photons. C_2 represents the illuminated solar cell array cover glass (quartz). A smaller parallel and also illuminated capacitor, C_3 (Kapton), was included in order to permit introduction of a nonlinear element from the reference node to the zero potential node at infinity. The values of these capacitors, C_1, C_2, C_3, were 4.5 μF, 2.4 μF, and 0.30 μF, respectively, for both winter and equinox seasons. During the summer season a part of the "dark" capacitance, C_1, becomes illuminated, and the values of C_1, C_2, C_3 become 3.0 μF, 2.0 μF and 1.8 μF, respectively. The leakage resistances were adjusted to correspond to the appropriate dielectric surface areas. In a more generalized analysis, where many more capacitive elements

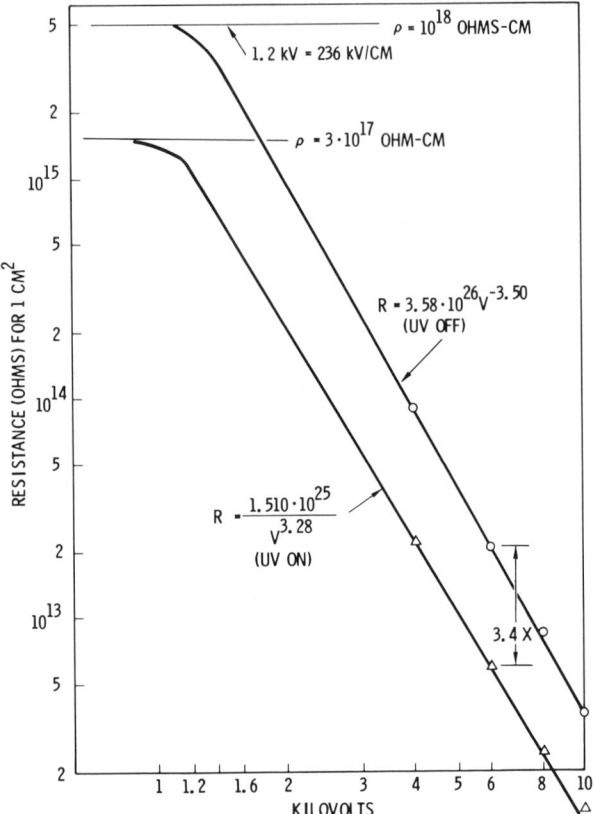

Fig. 6 Voltage and uv sensitivity of 2 mil Kapton resistance (data from J. M. Sellen, Jr.).

are to be included, a more systematic procedure should be used in which the seasonal and diurnal variations are incorporated only into the definitions of the current sources.

Current Sources

The various charging and discharging mechanisms constitute the forcing functions for the spacecraft charging model. In every case, electron and proton charging, photoemission, and secondary emission, the source is not representable as a simple linear function of potential. Because of the nonlinear nature of these current sources, as well as of the bulk conductivity, the analysis was carried out with a computer. The functional equations which relate currents to potentials are shown in Fig. 5. These equations are derived by J. Vogl in Appendix 2 of

SPACECRAFT POTENTIALS IN A SUBSTORM ENVIRONMENT

Ref. 4. Maxwellian temperatures, T, are expressed in terms of equivalent voltages, actually kT/e, in the equations for the currents. Also listed in Fig. 5 are the assumed parameters which define the substorm plasma, and the photoemission and secondary emission characteristics for dielectrics and metals.

Figure 7 shows the responses obtained with a step-function forcing function for the modified and unmodified spacecraft configurations. It shows the reduction in stress (V_0-V_1), achieved during the winter season by allowing the structure potential V_0 to follow V_1. In Fig. 7 and all subsequent figures showing surface potentials, sunlit dielectric surfaces, V_2, are at zero volts.

Figure 8 shows what happens when the substorm is turned off at the time (t ≅ 1200 sec), forming a "boxcar forcing function, when the stress has reached its maximum value in the modified configuration. It shows that in the response following turnoff, V_0 rises to zero by photoemission in about 6000 sec. For the case shown in Fig. 8 (T_e = 3 kv, T_p = 0.1 kv, J_e = 0.06 na/cm^2, J_p = 0), the voltage stress, V_0-V_1, actually continues to increase even though the initial step function con-

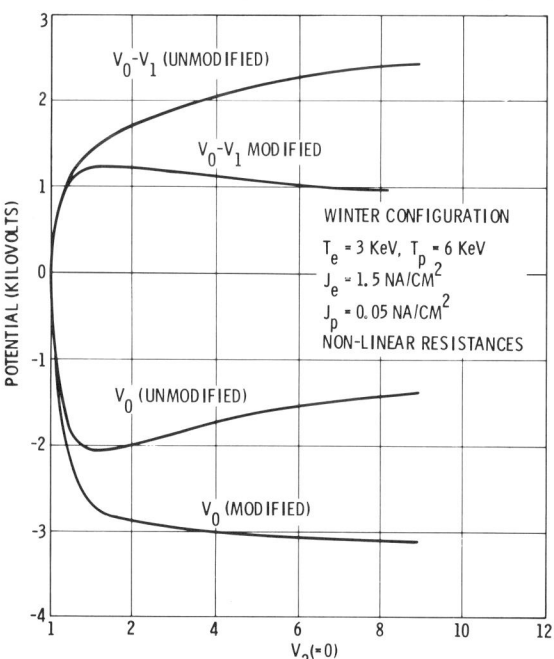

Fig. 7 Responses of modified and unmodified spacecraft configurations to step function forcing function.

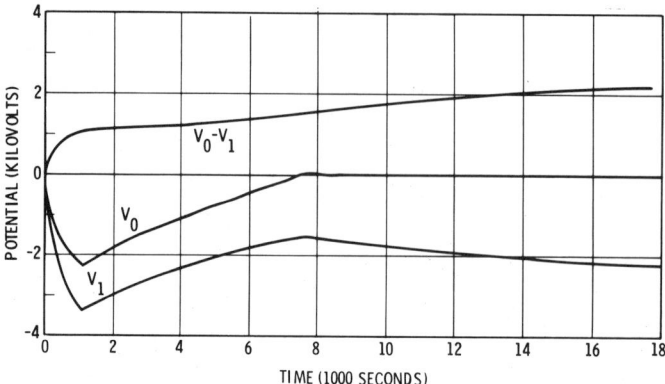

Fig. 8 Boxcar forcing function responses (turn-off J_e = .06 na/cm^2).

ditions (T_e = 3 kv, J_p = 6 kv, J_e = 1.5 na/cm^2, J_p = 0.0495 na/cm^2 have been turned off.

In Fig. 9 the turnoff conditions have been changed by reducing J_e from 0.06 na/cm^2 to 0.01 na/cm^2 and maintaining the other plasma parameters as in Fig. 8. The decay time for the voltage across the dark capacitor, V_0-V_1, is greater than one day and corresponds to the 2.65·10^5 sec or 3 days time constant for the thermal blanket material. The nonlinear reduction in leakage resistance with voltage stress above a 1.2 kv threshold has been included in the calculations for Fig. 7-9. The long time constant evidenced in Fig. 9 is obtained because V_0-V_1 is below the 1.2 kv threshold. An initial voltage greater than 1.2 kv would have caused a rapid decrease in V_0-V_1 to the threshold; then the much slower decay shown would have taken over.

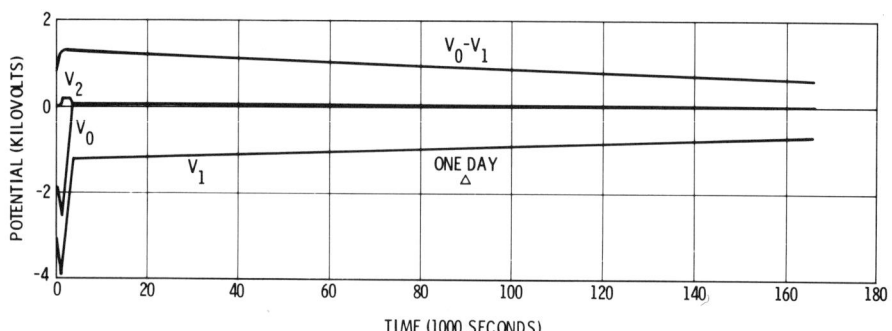

Fig. 9 Boxcar forcing function (turnoff J_e = .01 na/cm^2).

SPACECRAFT POTENTIALS IN A SUBSTORM ENVIRONMENT 113

Time Dependent Forcing Functions as Observed on ATS-5

The response of the spacecraft charging model to artificial forcing functions such as step functions and boxcar functions provides a good insight into the basic cause-and-effect relationships. Simulation of actual time-dependent in-orbit environmental conditions provides additional information which is also important. Figures 10-12 show data from ATS-5 on the environmental parameters for the first three days of 1972. It may be noted that Jan. 2nd is a more disturbed day, in that the current densities are much larger on that day than on the days before and after. The fluxes of currents also start earlier, at about 2100 hr local time, and last longer, until about 0900. On the other two days, the fluxes are confined pretty much to the midnight-to-dawn time sector.

The Maxwell-Boltzman temperatures, however, do not show any striking difference for all three days. If anything, a lowering of both ion and electron temperatures by a factor of about two during the midnight-dawn sector may be noted on all three days. It may also be noted that the ion temperatures are roughly twice that of the electrons. The assumption of plasma neutrality

$$\frac{J_e}{J_p} = \left(\frac{m_p T_e}{m_e T_p}\right)^{1/2} = 30.33 \text{ since } \frac{m_p}{m_e} = 1840 \text{ and } \frac{T_e}{T_p} = 1/2$$

seems to be roughly true when the electron flux is high (midnight/dawn). At the other parts of the day, however, the positive ion flux appears to be much greater. The explanation for this discrepancy probably lies in a combination of factors, such as the assumed Maxwell-Boltzmann energy distribution and the experimental energy detection range.

Figures 13-15 show spacecraft charge-up responses corresponding to the three days of observational data from ATS-5 given in Figs. 10-12, Jan. 1-3 of 1972. The same modified spacecraft configuration (Mod 2) as for Figs. 8 and 9 was used. However, the leakage resistance of the dark unilluminated capacitor was reduced by an order of magnitude ($\rho = 10^{17}$ Ω-cm instead of 10^{18} Ω-cm), and its nonlinear behavior was not included. The number of possible combinations of spacecraft configurations and materials parameters that could be run is very large, and only a few of them are presented here.

A number of observations may be made from the responses shown in Figs. 13-15. The first is that the threshold at which

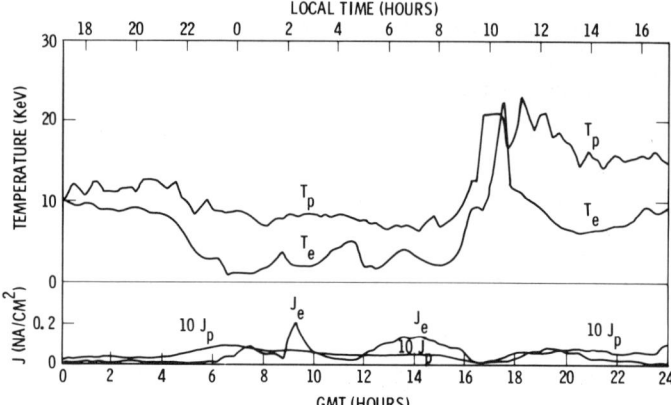

Fig. 10 ATS-5 environmental data for Jan. 1, 1972.

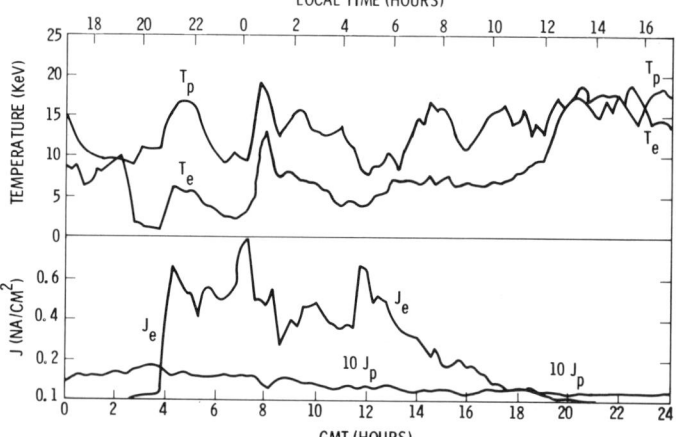

Fig. 11 ATS-5 environmental data for Jan. 2, 1972.

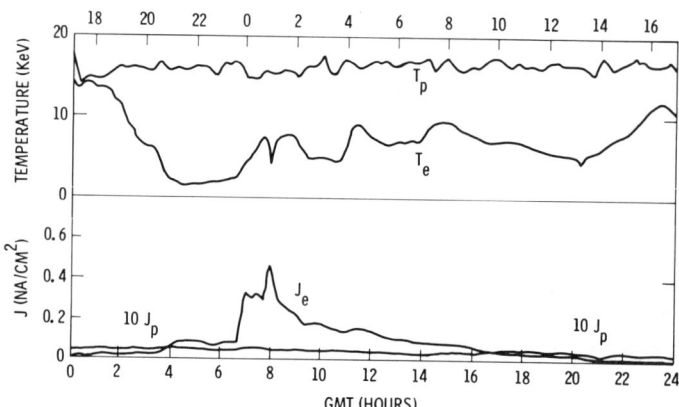

Fig. 12 ATS-5 environmental data for Jan. 3, 1972.

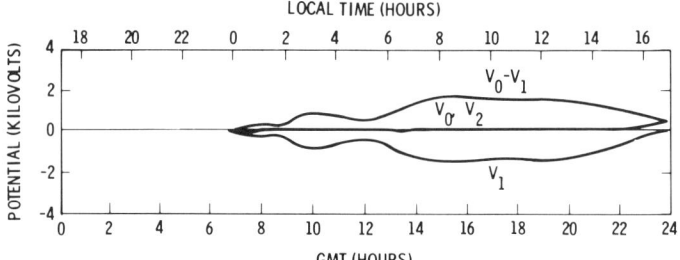

Fig. 13 Spacecraft potentials in response to the Jan. 1, 1972 environment.

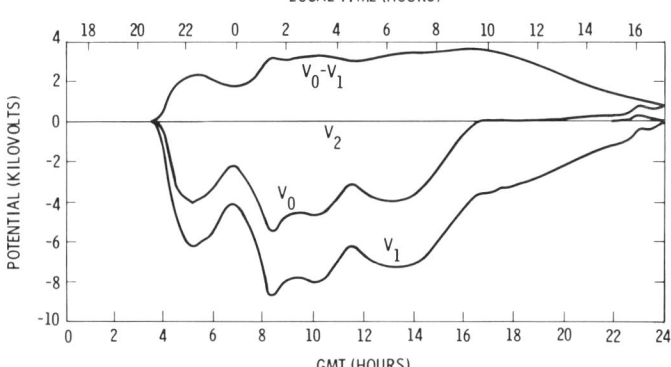

Fig. 14 Spacecraft potentials in response to the Jan. 2, 1972 environment.

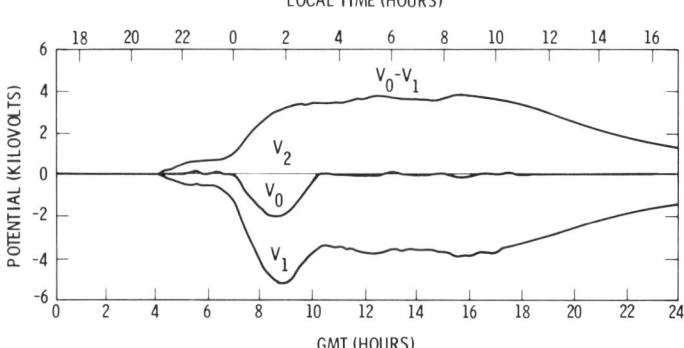

Fig. 15 Spacecraft potentials in response to the Jan. 3, 1972 environment.

charging begins to take place does require the enhanced electron flux which appears near midnight. All of the response curves verify that the potentials do not change significantly within time periods shorter than 10 or 15 min. It may be noted that the charge or stress on the dark capacitor (V_0-V_1) on the

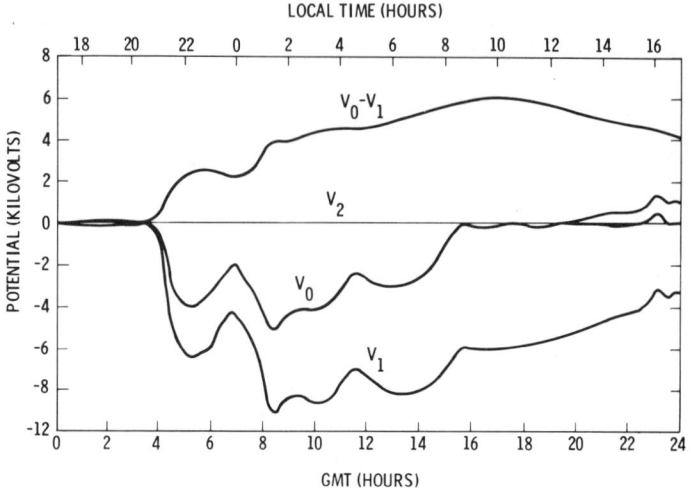

Fig. 16 Response for Jan. 2, 1972 with $\rho = 10^{18}$ Ω-cm.

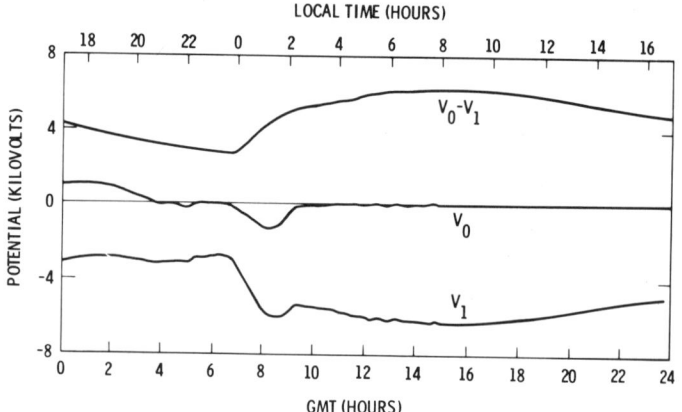

Fig. 17 Response for Jan. 3, 1972 with $\rho = 10^{18}$ Ω-cm.

third day is nearly the same as on the second day, even though the fluxes are less than half as large. On the first day, the maximum stress is about one-half that on the other two days. The integrated flux on the first day is roughly a half of that on the third day, and that for the second day is roughly three times that on the third day.

The difference between the second and third days is that the structure potential, V_0, is "unclamped" on the second day. Presumably, if it had been in the unmodified configuration, V_0 would have been clamped and the V_0-V_1 stress would have been much larger. Note that the maximum "dark" potential, V_1, on the second day is about twice that on the third day.

Figures 16 and 17 for Jan. 2 and 3 demonstrate the effects of assuming a dielectric resistivity for the "dark" capacitor material of 10^{18} Ω-cm rather than 10^{17} Ω-cm as was done for the calculated responses of Figs. 13-15. Compared to the latter, the V_0-V_1 stress attains values about 50% greater on Jan. 2, and has decayed only to about one-half of the maximum by local midnight of the next day, when charge-up commences again. On Jan. 3 the maximum stress is again about 50% higher than on Jan. 1 and has decayed only by about 15% at GMT midnight of that day. Thus, if the higher resistivity of 10^{18} Ω-cm (the nominal manufacturer's value) is appropriate, a day-to-day buildup of stress is a possible consequence.

Figures 18 and 19, both for Jan. 1, 1972, demonstrate the effects of positive ions on the character of the charge-up. The difference for the two figures is that the ion current density was reduced by 99% in the calculations for Fig. 18, but was not for Fig. 19. For both cases, the dark capacitor material resistivity was taken to be 10^{18} Ω-cm. This accounts for the difference between these responses and those of Fig.

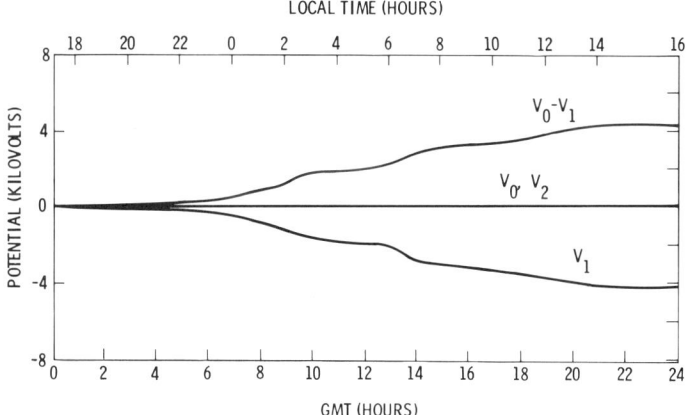

Fig. 18 Response for Jan. 1, 1972 with J_p reduced by 99% ($\rho = 10^{18}$ Ω-cm).

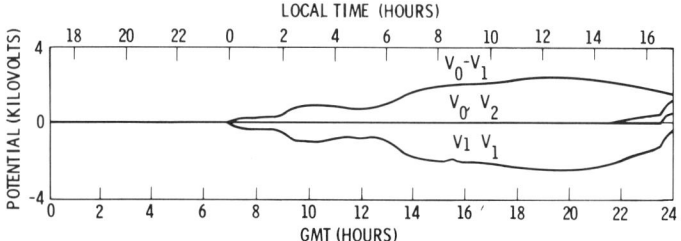

Fig. 19 Response for Jan. 1, 1972 with 100% J_p ($\rho = 10^{18}$ Ω-cm).

13 for the same day. Comparison of Figs. 18 and 19 shows that positive ions do limit the charge-up to negative potentials by electrons. If, for example, only the positive ions were field aligned and of much smaller fluxes, this would be a situation equivalent to that of Fig. 18, in which only electrons were present. However, the ATS-5 plasma measurements indicate that the electron number density did not exceed the ion density by a large amount for any length of time (S. DeForest, private communication).

Experimentally, as noted before, an excess of ions was observed; especially when ATS-5 was outside the midnight/dawn sector. At these times, since one would expect charge neutrality, the plasma sheath should have a large surplus of low energy electrons with energy below the instrument detection threshold. These electrons in turn should prevent charging to appreciably large positive potentials. In a number of the response curves shown, positive potentials of several hundred volts may be noted near local dusk. These are because the observational forcing function data have not included these low-energy electrons, and the positive potentials therefore should not be considered to be real. In the computational program, we have shown that providing these electrons by exponentially increasing electron currents and decreasing ion currents for positive surface potentials eliminates these large positive potentials.

A materials characteristic of Kapton is that its resistivity decreases by four orders of magnitude to 10^{14} Ω-cm when the temperature is increased from 25°C to 200°C. Qualitatively, this lowering of resistivity should affect illuminated dielectric surfaces, but not those that are not illuminated. This would tend to lower the effectiveness of modifications which attempt to "unclamp" the structure potential. In general, this type of thermal behavior tends to decrease stresses in illuminated dielectrics, and to increase stresses in those which are not illuminated. The relatively fewer number of events in the summer as compared to the winter months, even though stresses in the forward closure are only slightly less according to the calculations, partially may be due to this thermal effect on resistivity. Other factors also must be considered, however, such as the relationship between stress levels and voltage breakdown levels, and the effect of the difference in stress polarity on the susceptibility of a given electronic circuit to arc discharges caused by that stress.

Conclusions and Recommendations

The results of the spacecraft charging model analysis with both artificial (step function, boxcar) and real observational-type forcing functions show that the potentials are strongly dependent on the materials properties (resistivity, photoemission, and secondary emission); configuration; and orientation of each particular spacecraft. The analytical methodology used involves gross simplifications, and much further work remains to be done. The demonstrated consistency of the results with the observations (some predicted in advance) with the occurrences of in-flight anomalies supports the validity of the analysis. Specific results of the analysis are as follows:

1) Spacecraft differential charging time constants are on order of tens of minutes.

2) Discharge times may be on the order of many hours.

3) Buildup of potential differences may occur over a number of successive substorms.

The conclusions and recommendations resulting from this study are as follows:

1) More data on materials characteristics are required from laboratory and in-flight experiments.

2) More data to define the geosynchronous environmental parameters are required.

3) Improved analytical techniques for applying the above data to predict spacecraft potentials should be developed.

The motivation for this study was to develop methods for designing spacecraft that are able to withstand the geosynchronous orbit environment. Prediction of the potentials developed by spacecraft charging constitutes an important factor in the design. In addition to the aforementioned measures to obtain better design, it is recommended that every spacecraft be flown in geosynchronous orbits be tested before launch to verify its immunity to the environment. A minimal housekeeping type diagnostic should be incorporated into every such spacecraft to monitor its state-of-health in regard to this aspect of its operational capability. The monitors should be able to detect charging at several locations, as well as the noise injected into critical cable harnesses by arc discharges. In the event of the occurrence of abnormal behavior, these monitors will provide important diagnostic information.

References

[1] Rosen, A., Fredricks, R. W., Inouye, G. T., Sanders, N. L., Scarf, F. L., Greenstadt, E. W., Vogl, J. L., and Sellen, J.

M. Jr., "Final Report RGA Analysis: Findings Regarding Correlation of Satellite Anomalies with Magnetospheric Substorms, and Laboratory Test Results," TRW Defense & Space Systems, Redondo Beach, Calif., Rept. 09670-7020-R0-00, Aug. 1, 1972.

[2] "Final Report TDAL Gain State Analysis," TRW Defense & Space Systems, Redondo Beach, Calif., Rept. 09670-7040-RU-00, Oct. 18, 1973.

[3] "Final Technical Report Program 777 Anomaly Investigation for Satellites 9433 and 9434," Vol. II, TRW Defense & Space Systems, Redondo Beach, Calif., Rept. 09670 RFP 050-01, March 22, 1974.

[4] Inouye, G. T., Vogl, J. L., Sellen, J. M. Jr., DeForest, S. E., and Rosen, A., "Final Report Spacecraft Charging Analysis: A Study and Analysis of the DSCS II Spacecraft Orbital Charging Phenomena," TRW Defense & Space Systems, Redondo Beach, Calif., Aug. 16, 1974.

[5] "Final Report, Spacecraft Charging Analysis II, Studies and Analysis of the Modified DSCS II Flights 5, 6 Configuration," TRW Defense & Space Systems, Redondo Beach, Calif., Rept. 9670-REP-052-01, March 31, 1975.

[6] Rosen, A., "Spacecraft Charging: Environment Induced Anomalies," AIAA Paper 75-91, Jan. 1975, Pasadena, Calif.

[7] DeForest, S. E., "Spacecraft Charging at Synchronous Orbit," Journal of Geophysical Research, Vol. 77, Feb. 1972, pp. 651-659.

[8] DeForest, S. E., "Electrostatic Potentials Developed by ATS-5," Photon and Particle Interaction with Surfaces in Space, edited by R. J. L. Grard, D. Reidel Publishing Co., Dordrecht-Holland, 1973, pp. 263-276.

[9] Inouye, G. T., "Spacecraft Charging Model," AIAA Paper 75-255, Jan. 1975, Pasadena, Calif.; also Journal of Spacecraft and Rockets, Vol. 12, No. 10, Oct. 1975, pp. 613-620.

[10] Hoffmaster, D. and Sellen, J. M. Jr., "Electron Swarm Tunnel Measurements of Kapton Bulk Resistivity at High Electric Stress Levels," TRW Defense & Space Systems, Redondo Beach, Calif., Rept. 4351.3.74-39, Sept. 5, 1974.

SIMULATION OF THE PLASMA SHEATH SURROUNDING
A CHARGED SPACECRAFT

P. L. Rothwell,[*] A. G. Rubin,[+] A. L. Pavel,[‡] and
L. Katz[§]

Air Force Geophysics Laboratory
Hanscom Air Force Base, Mass.

Abstract

A self-consistent computer code that simulates the particle-field interaction in the sheath surrounding a charged, spherically symmetric body is described. Monte Carlo techniques are used to generate the appropriate Maxwellian velocity and angular momentum distributions. Starting with an ambient plasma, the sheath structure is traced in a time-dependent manner to its steady-state condition for a given probe (satellite) potential and plasma parameters. Good agreement is shown with the orbit-limited Langmuir theory where appropriate. The expected photosheath density distribution about a charged spacecraft is developed. Finally, it is shown that the floating potential depends on the electron/ion temperature ratio as well as the secondary emission properties of the satellite surface material.

Introduction

Spacecraft-charging now is believed to be responsible for the anomalous behavior of synchronous orbit satellites

---

Presented as Papers SA64 and SA65 at the American Geophysical Union 56th Annual Meeting, Washington, D.C., June 16-19, 1975. We should like to express our appreciation to Lee W. Parker, whose opinions and advice during various stages of this work are greatly appreciated.
[*]Staff Physicist, Energetic Particles Branch.
[+]Staff Physicist, Energetic Particles Branch.
[‡]Staff Physicist, Energetic Particles Branch.
[§]Chief, Energetic Particles Branch.

observed during magnetic substorms.[1-4] A complete model for this phenomenon must include satellite-sheath interactions, including satellite-return currents due to differential charging. One approach is to model the various satellite surfaces and to treat the sheath as an external source.[5] A complementary approach, taken in the present work, is to treat the sheath as an integral part of the satellite-environment interaction. Starting with a simple satellite geometry, one gradually makes the geometry more complex and thus constructs a more realistic sheath model. Both approaches lead to a complete model, but from different starting points.

High-time resolution particle detectors[6] will be flown onboard the SCATHA (spacecraft charging at high altitudes) satellite scheduled to be launched in 1978. In order to interpret the data properly, it will be necessary to determine the time-dependent behavior of the sheath. For this purpose, we have constructed a computer simulation code that models the particle-sheath interaction and, in particular, highlights the time-dependent behavior. This code treats the spherically symmetric case with an ambient plasma of arbitrary density and temperature. The code also generates photo and secondary electrons, and can simulate emission from an ion/electron source as well. By turning the aforementioned processes on and off under various assumed conditions, one can conduct time-dependent computer simulation experiments. Because the satellite velocity at synchronous altitude is negligible compared with the mean thermal velocity of kilovolt plasmas, drift velocities have not been considered here.

Computer Code

The present approach follows that of Albers,[7] and traces simultaneously the trajectories of a few thousand computer electrons and ions constrained to obey Gauss' law, conservation of angular momentum, and Newton's equations of motion. Injection of computer particles at the sheath boundary and their subsequent absorption at the satellite simulates current flow in and through the physical sheath. The sheath structure can be followed in time to its steady-state condition.

To use the program under a wide range of plasma conditions, we assign weights to the individual computer particles. For a constant ambient density, the number of particles in a given spherical shell is obviously directly proportional to the volume of that shell. If one were to weight the initial particle distribution in a position-independent manner, a large number of computer particles would be found near the sheath boundary (r = R_B) and very few near the satellite. The poor sampling near the satellite would lead to unacceptable noise levels. We circumvent this difficulty by noting that the particle population existing at any position and time is composed of particles that started at time zero in various volume elements. By substituting for each class of particles initially in the same volume element a smaller number of weighted computer particles, one can simulate systematically the particle flow between volume elements over time. Each of the computer particles (electrons or ions) is, therefore, assigned a weight:

$$W_j = n_o V_i^- / \bar{N} \quad (1)$$

where n_o is the ambient plasma density, V_i^- is the volume element of the ith spherical shell in which the jth particle originated, and N is the number of assigned computer particles per spherical shell. In the present study, N ∼ 200-300.

The weighted computer particles are distributed uniformly throughout the associated spherical shell with a Maxwellian velocity distribution. The Maxwellian velocity distribution for V_r is given by

$$f(V_r) = (m/2\pi kT)^{1/2} e^{-\frac{1}{2}mV_r^2/kT} \quad (2)$$

Setting

$$y = (m/2kT)^{1/2} V_r = \frac{1}{\sqrt{2}} V_r / V_{TE}, \quad dy = (m/2kT)^{\frac{1}{2}} dV_r \quad (3)$$

where V_{TE} is the root-mean-square thermal electron velocity.

Then

$$f(V_r) dV_r = f(y) dy = e^{-y^2} dy / \sqrt{\pi} \quad (4)$$

Therefore,

$$P = \int_0^y f(y') dy' / \int_0^\infty f(y') dy' \equiv erf(y) \quad (5)$$

where P is a uniform distribution of random numbers between zero and one. Each random number P now is associated with a specific velocity V_r with a weight of $f(V_r)$.

The angular momentum distribution is found by generating a set of random numbers for the ∅ and θ (transverse) components using Eq. (5). The angular momentum of each computer particle is given by

$$J^2 = m^2 r^2 (V_\varphi^2 + V_\theta^2) \qquad (6)$$

where r is the initial particle position.

The particles constituting this initially Maxwellian plasma then are moved in response to the self-consistent force fields acting upon them. The time-step size (Δt) is chosen such that an electron moving at a velocity V_{TE} will move some small fraction of a shell width. Δt then is lowered systematically until further lowering does not affect the results.

During each time step, the computer particles are sorted according to their spherical shell location. Each charge shell is incremented by $W_{j,i}$ for each proton and decremented by $W_{j,e}$ for each electron. Gauss' Law then is used to calculate the electric field using the computer generated total charge. Thus, the time evolution of the sheath electric field and potential is traced.

Following Bernstein and Rabinowitz,[8] we assume the potential to drop off as r^{-2} for $r > R_B$. This leads to the boundary condition $\phi_B = R_B E_B/2$, where E_B is the value of electric field on the boundary ($r = R_B$).

The potential at each shell boundary is then

$$\phi(r_i) = -\int_{R_B}^{r_i} E \cdot dr + \phi_B = \sum_{j=i}^{N+1} E_j \Delta r_j + \phi_B \qquad (7)$$

where the integral has been approximated by a summation over shells.

The probe potential $\phi_P = \phi(r_p)$ either can be fixed at some value ϕ_P or can be left floating at a potential, dependent on the number of ambient ions and electrons collected by the probe. For spacecraft-charging, the floating potential case is the appropriate one. The fixed potential case is treated here for comparison with past work. During each time step, ions and electrons leave and enter the sheath boundary. The number of random electrons entering are calculated exactly by the expression

$$N_E = 2 \sqrt{2\pi} V_{TE} R_B^2 n_o \Delta t \qquad (8)$$

with a similar expression for N_i.

In general, N_E is too large to be handled as individual particles, and the injected computer particles are weighted in a manner similar to that described for the ambient plasma. In each time step, $\bar{N}$ computer particles are injected at R_B with a Maxwellian velocity distribution in the negative r direction and with an appropriate angular momentum distribution. Because of the preceding considerations, $\bar{N}$, N_E, and W_N are related by

$$N_E = W_N \cdot \bar{N} \tag{9}$$

where $W_N = n_o \, \mathcal{V}_N/\bar{N}$ is the weight corresponding to the last spherical shell, which is bounded by R_B. If W_N is too large, $\bar{N}$ is small and the statistical representation at the boundary is poor. On the other hand, if W_N is too small, $\bar{N}$ is too large and the computer storage capacity is exceeded quickly. Thus, one must choose W_N such that $\bar{N}$ lies between these two extremes. To handle this problem, a parameter f was introduced such that $\bar{N} = fN$. Combining this with expression (8) leads to

$$\mathcal{V}_N = 2 \sqrt{2\pi} \, V_{TE} \, R_B^2 \, \Delta t/f \tag{10}$$

The volume of the last spherical shell is, therefore, set by the choice of f.

In the present study, we are concerned with the thick-sheath approximation in which the satellite potential is the floating potential. In the absence of secondaries, the satellite floating potential depends on the ratio T_e/T_i. For the cases studied here, this potential lies near $2.6 \, kT_e$. Within a few satellite radii, the potential drops to a fraction of the electron thermal energy. Under these conditions, it is not possible to delineate precisely a sheath boundary as in the thin-sheath case. We can define only a region beyond which the potential is less than the kT_e and where $\emptyset$ asymptotically approaches zero. For the purposes of this work, we take the sheath boundary to be where $\emptyset(r) \sim kT_e$.

In handling presheath acceleration, it is assumed that a Maxwellian velocity distribution exists at the boundary modified only by the Boltzmann factor. This is realistic in that one expects particle velocities to be random at large distances. The Boltzmann factor is $\exp(\pm e\emptyset_B/kT_{e,i})$. The injected particle energy in the -r direction is incremented by $\pm e\emptyset_B$. The plus sign corresponds to the attracted species and the negative sign to the repelled species.

The Boltzmann factor causes the program results to be insensitive to the boundary position. At smaller R_B, more

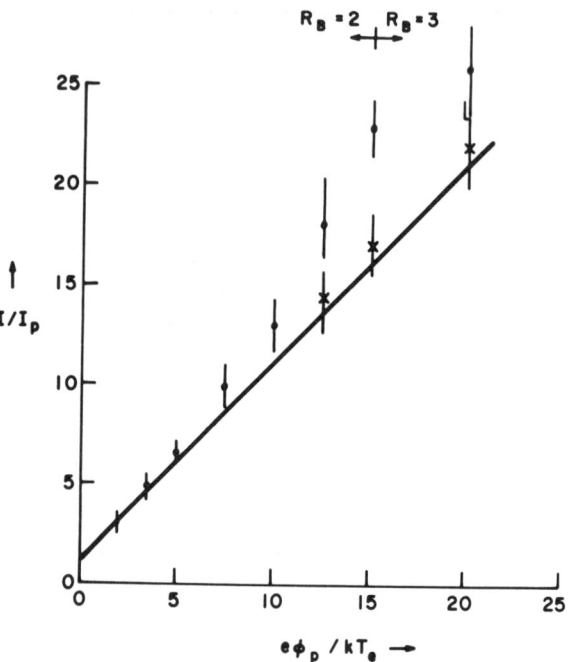

Fig. 1 Comparison of the simulation code results with the orbit-limited Langmuir theory. I_p is the ambient plasma current to the satellite at zero voltage. R_B refers to the sheath boundary in multiples of λ_D.

directional particles are injected because of a higher ϕ_B. At larger R_B, fewer particles with lower directionality are injected. Thus, the program simulates the natural particle behavior at different boundary positions.

Results

It is shown now that the aforementioned computer code yields agreement with the orbit-limited Langmuir probe theory.[9] In addition, the versatility of the approach is demonstrated by the generation of a photosheath about the satellite and the treatment of time-dependent charge buildup. Finally, the dependence of the floating potential on the electron/ion temperature ratio is determined.

SIMULATION OF THE PLASMA SHEATH 127

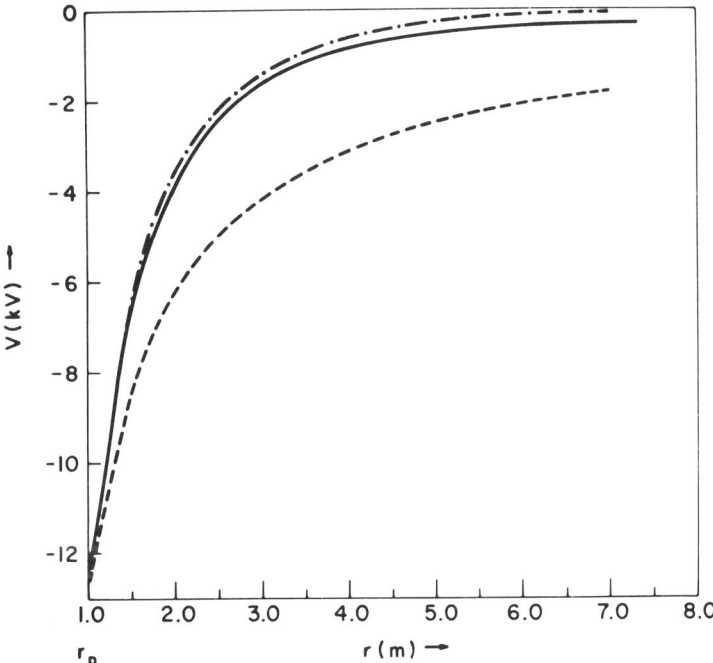

Fig. 2 Potential distribution generated by the computer code compared with a debye-shaped potential.

Comparison with Langmuir Theory

Figure 1 shows comparisons of the code results with the Langmuir theory. The error bars are voltage fluctuations due to the finite number of computer particles as just described. Good agreement is seen except at higher voltages, where the computer code predicts higher currents. This is because the code includes presheath acceleration, whereas the Langmuir theory does not. If the presheath acceleration is removed from the calculation, agreement with the Langmuir theory is obtained everywhere.

A further check is made to see if the program generates a debye-shaped potential in the sheath. To test this case, we set the plasma parameters so that λ_D is relatively small.

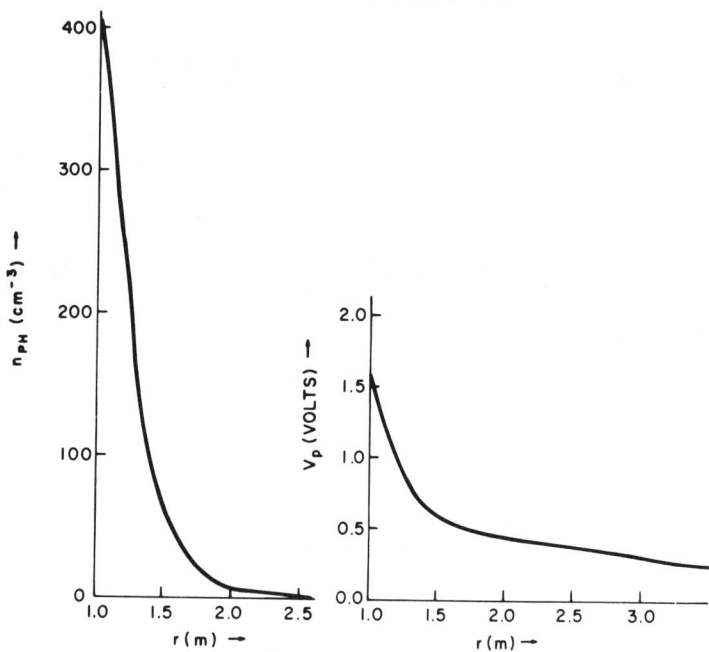

Fig. 3 Expected photoelectron sheath about a $r_p = 1$ m spacecraft. The actual sheath will be very dependent on the different material properties of the spacecraft surface.

From Fig. 2, one sees that the obtained potential distribution is approximated closely by the debye potential. A detailed computer calculation using the PARKLAF[10] program showed similar agreement. This figure also shows the $1/r$ potential for comparison. These detailed agreements act to reassure us of the results obtained by our simulation. It also shows that the program correctly simulates the thin-sheath case as λ_D is made smaller.

Photosheath

A photoelectron sheath is created by the emission of electrons from the satellite surface. Because the satellite has a floating potential, a positive charge builds up which attracts the emitted electrons back to the surface. When the total current flow to the satellite is zero, equilibrium is

SIMULATION OF THE PLASMA SHEATH

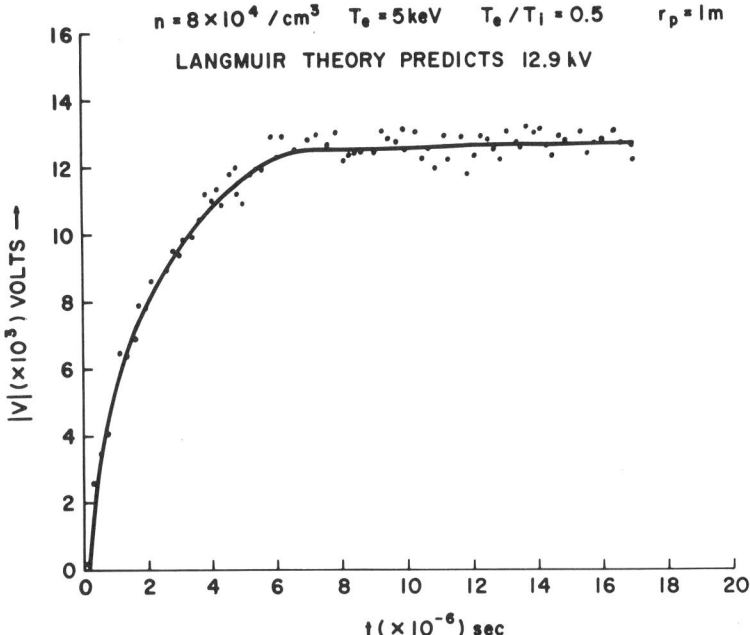

Fig. 4 Time-dependent voltage buildup on a r_p = 1 m spacecraft; t scales inversely with n.

reached. This is simulated by the emission of a number of weighted computer particles whose trajectories are traced. The plasma parameters are adjusted so that the ambient current is much less than the photocurrent. The results are shown in Fig. 3 and are consistent with those obtained by Schroder.[11]

Time Dependence

The simulation method also delineates the time-dependent behavior. In particular, we are interested in determining the time constant for a satellite to reach its steady-state floating potential for a given plasma temperature and density. Unfortunately, the plasma parameter values measured in space ($n_o \sim 1/cm^3$, $T_e \sim 5$ kev, $\omega_p \sim 40$ khz) imply a time constant that would require an excessively large amount of computer time. To circumvent this difficulty, we note that in the Langmuir theory the floating potential for a spherical probe is independent of the ambient density n_o. Since the current density is directly proportional to n_o the rate of charge buildup also is proportional to n_o. Therefore, the time scale for the satellite to accumulate a given amount of charge is inversely proportional to n_o.

Figure 4 shows the simulated time response for $n_o = 8 \times 10^4/cm^3$, $T_e = 5$ kev, and $T_e/T_i = 0.5$. The satellite reaches equilibrium in about 6 μsec. The dots represent actual output from the code, and the spread in data represents the noise inherent in the simulation. It is noted that the spread in the data includes the value predicted by the Langmuir theory. Multiplying the time scale by n_o, we find that the expected time constant for satellite charge buildup at $n_o = 1/cm^3$ is 0.4 sec. This shows that the rapid-scan instrumentation to be flown aboard SCATHA[6] will be able to resolve the expected time changes in the measured flux.

Langmuir Theory Floating Potential

The floating potential for a spherical probe can be obtained directly from Langmuir's theory.[12] The repelled particles, which in this case are electrons, are assumed to be in equilibrium about the spacecraft according to the Boltzmann formula

$$n(r) = n_o \exp[-e\phi(r)/kT_e] \qquad (11)$$

where $\phi(r)$ is the electric potential.

The ion current to the satellite is given by Chen[12]:

$$J_i = 4\pi r_p^2 \tfrac{1}{2} n_o (2kT_i/\pi m_i)^{\tfrac{1}{2}} F \qquad (12)$$

where, in the limit $r_p \ll \lambda_D$, $F \sim 1 + e\phi_p/kT_i$; $\phi_p \equiv \phi(r = r_p)$. The electron current to the satellite is

$$J_e = 4\pi r_p^2 \tfrac{1}{2} n_o \frac{(2kT_e)^{\tfrac{1}{2}}}{\pi m_e} e^{-e\phi_p/kT_e} \qquad (13)$$

Setting $J_i = J_e (1 - \alpha)$, where α is the coefficient of secondary emission, one obtains the relation

$$(1-\alpha) m_i T_e/m_e T_i)^{1/2} \exp(e\phi_p/kT_e) = 1 + e\phi_p/kT_i \qquad (14)$$

Equation (14) can be expressed as a quadratic in $(T_e/T_i)^{\tfrac{1}{2}}$. The solutions are shown in Fig. 5, together with results from the simulation program. The important implication of these results is that the expected floating potential is very dependent on the presence of low-temperature protons, as well as the secondary emission properties of the satellite surface. Figure 5 covers the range of parameter values expected at geosynchronous altitude.

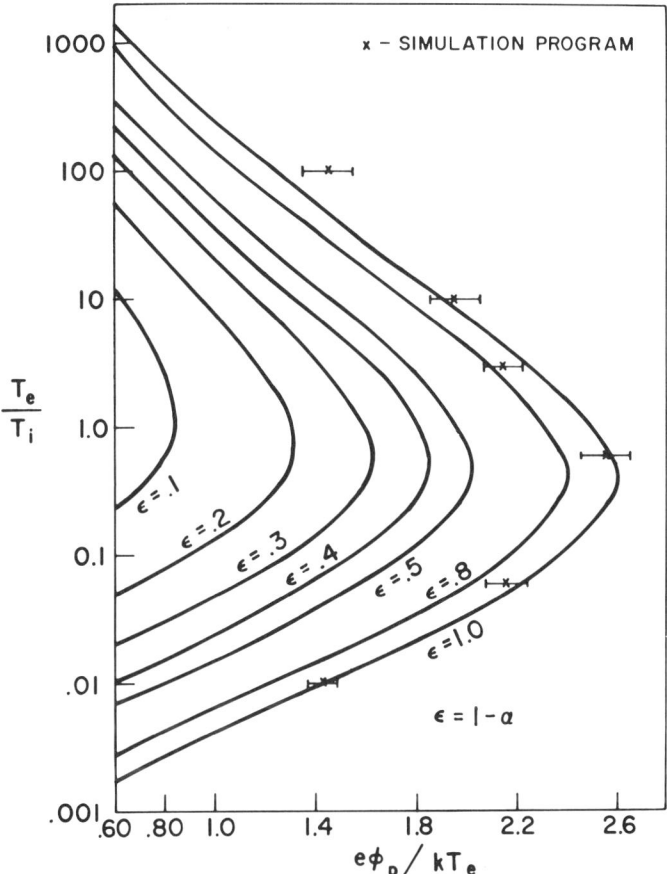

Fig. 5 Floating potential vs the electron/ion temperature ratio. ϵ is equal to $1-\alpha$, the coefficient of secondary emission. The horizontal bars correspond to the code results, assuming no secondary emission.

Summary and Conclusions

The current voltage characteristics of probes in collisionless plasmas have been studied extensively by Laframboise.[13] Employing the Bernstein-Rabinowitz theory formulation, he obtained numerical solutions using Maxwellian electron and ion distributions. The present work treats the probe in a collisionless plasma by computer simulation. The advantage of the present treatment is that it can be extended to more complicated situations. In addition to time-dependent behavior, photoemission and secondary emission are included. It is possible to treat the isotropic emission of ion and electron beams from the probe surface also. It is concluded from the present work that a spacecraft at synchronous orbit would charge up in approximately 0.4 sec, which is within the contraints of the rapid-scan instrumentation to be flown on SCATHA.[6] It is concluded further that the electron/ion temperature ratio of the ambient plasma must be known before the secondary emission properties of the spacecraft surface can be deduced from the floating potential. Future research is aimed at deriving further checks on the present code and extending it to higher dimensions in order to obtain more realistic results.

References

[1] Rosen, A., Fredericks, R. W., Inouye, G. T., Saunders, N. L., Scarf, F. L., Greenstadt, W. W., Vogl, J. L., and Sellen, J. M., "RGA Analysis: Findings Regarding Correlation of Satellite Anomalies with Magnetospheric Substorms and Laboratory Test Results," Rept. 09670-7020-R0-00, Aug. 1972, TRW Systems Group, Redondo Beach, Calif.

[2] DeForest, S. E., "Spacecraft Charging at Synchronous Orbit," *Journal of Geophysical Research*, Vol. 77, No. 4, Feb. 1972, pp. 651-659.

[3] McPherson, D. A., Cauffman, D. P., and Schober, W., "Spacecraft-Charging at High Altitudes; the SCATHA Satellite Program," AIAA Paper 75-92, Jan. 1975, Pasadena, Calif.

[4] Pike, C. P. and Bunn, M. H., "A Correlation Study Relating Spacecraft Anomalies to Environmental Data," published elsewhere in this volume.

[5] Inouye, G. T., "Spacecraft Potentials in a Substorm Environment," published elsewhere in this volume.

[6] Sellers, B., Hanser, F. A., Morel, P. R., Hunerwadel, J. L., Pavel, A. L., Katz, L., and Rothwell, P. L., "Design and Calibration of a High Time Resolution Spectrometer for 0.05 to 500 kev Electrons and Protons," published elsewhere in this volume.

[7] Albers, N., "Computer Simulation of a Spherical Langmuir Probe," SU-IPR Rept. 499, 1973, Stanford University, Stanford, Calif.

[8] Bernstein, I. B., and Rabinowitz, I. N., "Theory of Electrostatic Probes in a Low-Density Plasma," *The Physics of Fluids*, Vol. 2, No. 2, March-April 1959, pp. 112-121.

[9] Mott-Smith, H. M. and Langmuir, I., "The Theory of Collectors in Gaseous Discharges," *Physical Review*, Vol. 28, Oct. 1926, pp. 727-763.

[10] Parker, L. W., private communication, Oct. 1975.

[11] Schroder, H., "Spherically Symmetric Model of the Photoelectron Sheath for Moderately Large Plasma Debye Lengths," *Photon and Particle Interactions with Surfaces in Space*, edited by R. J. L. Grard, D. Reidel Publishing Co., Dordrecht, Holland, 1973.

[12] Chen, F. F., "Electric Probes," *Plasma Diagnostic Techniques*, edited by R. H. Huddlestone and S. L. Leonard, Academic Press, New York, 1965.

[13] Laframboise, J. G., "Theory of Spherical and Cylindrical Langmuir Probes in a Collisionless, Maxwellian Plasma at Rest," University of Toronto Rept. 100, 1966, Toronto, Canada.

THEORY OF THE SPHERICALLY SYMMETRIC PHOTOELECTRON SHEATH AND COMPARISON WITH THE ATS-6 OBSERVATION OF A POTENTIAL BARRIER

E. C. Whipple Jr.[*]

NOAA Environmental Research Laboratories,
Boulder, Colo.

Abstract

Photoelectrons and secondary electrons recently have been observed to have been reflected back to the ATS-6 spacecraft from a potential barrier in the spacecraft vicinity. The presence of such a potential barrier in a photoelectron sheath has been theoretically predicted. In order to determine whether or not the observed barrier is due to the emitted electrons, a theory for a spherically symmetric sheath has been developed which includes the effects of both the emitted electrons and the ambient plasma electrons and ions. A thick sheath approximation based on an analysis of particle trajectories in phase space is obtained which is valid for large particle Debye lengths. A comparison with the ATS-6 data

---

Presented as Paper No. AS37 at the AIAA/AGU Spacecraft Charging by Magnetospheric Plasmas Conference, June 17, 1975; also published in The Journal of Geophysical Research, Vol. 81, pg. 601, 1976, copyrighted by The American Geophysical Union. This work was stimulated by the invitation of Carl McIlwain and Sherman DeForest to spend several weeks at the University of California in San Diego to study the ATS-6 data. I would like to thank them for many helpful discussions of this problem.

[*]Research Physicist, Aeronomy Laboratory; presently Research Physicist, Physics Department, University of California, San Diego.

shows that the observed barrier potentials are too large to be explained in this way.

Introduction

It is well known that photoelectrons emitted by spacecraft in sunlight can be important in determining the spacecraft potential if the ambient plasma density is low enough. It also is known that secondary electrons emitted from a spacecraft surface by the action of energetic primary particles can affect the interpretation of low-energy particle experiments,[1] especially in the magnetosphere and in the solar wind, where the environmental plasma is characterized by a low density but high energies.

A number of workers have discussed the effect of photoelectrons on the plasma sheath surrounding a spacecraft. Singer and Walker[2] calculated approximately the screening effect of photoelectrons on a positively charged spherical body. Medved[3] discussed the effects of secondary electrons as well as photoelectrons on a satellite sheath. Grard and Tunaley[4] compared the effects of monoenergetic, rectangular, and Maxwellian velocity distributions for the photoelectrons on a planar (one-dimensional) sheath, and concluded that beyond the plasmapause probe measurements probably would be affected by photoemission effects. Guernsey and Fu[5] and Fu[6] showed that if the photoelectrons dominate the space charge near the satellite surface it is possible for a potential minimum to develop in the sheath so that the potential distribution is nonmonotonic. They treated the idealized infinite plane photoemitting surface that is appropriate if the Debye length for the photoelectrons is small compared to satellite dimensions. More recently, Schroder[7] used the method of Laframboise[8] to obtain solutions for the potential distribution in a thick sheath about a sphere under conditions pertaining to the solar wind. He found a potential minimum of about -1 V with respect to the plasma, with an assumed positive spacecraft potential. Tunaley and Jones[9] treated approximately the spherically symmetric thick sheath, but without taking into account the effect of the ambient plasma. Since the Debye length for photoelectrons is on the order of 1 m or greater, it appears that a three-dimensional treatment is necessary to obtain a realistic description of a photoelectron sheath about the typical spacecraft.

We recently have described evidence from the ATS-6 spacecraft for the presence of photoelectrons and secondary electrons reflected from a potential barrier in the spacecraft vicinity.[10] In order to determine whether this potential barrier was a photoelectron (or secondary electron) effect of the kind predicted by Guernsey and Fu, or whether it was some other phenomenon, it was necessary to develop a theory for the photoelectron sheath which could be compared easily with the ATS-6 data. The purpose of this paper is to present this theory and to compare calculations using this theory with the experimental data.

Formulation of the Problem

We want to find the potential distribution in the sheath about a body that is emitting electrons. The body is immersed in a plasma, where it may be assumed that the ambient electron and ion density and temperatures are known. We are interested especially in the case where a potential barrier can develop. The problem is equivalent to finding simultaneous solutions for the Poisson equation and the collisionless Boltzmann (Vlasov) equations. The solution to the Vlasov equation is superficially trivial; since the problem is collisionless, the distribution function is a constant along any particle trajectory. However, what is not trivial is finding the demarcation in phase space between different types of trajectories. These boundaries in phase space determine the limits of the density integrals and are the crux of the problem. Once expressions for the particle densities have been obtained in terms of the local potential, Poisson's equation may be integrated numerically to obtain the potential as a function of radius.

In the general case, it is not possible to express the particle density in terms of the local potential. The reason for this is that the boundaries in phase space between the different types of trajectories depend upon the potential distribution elsewhere in the sheath and, in particular, upon the potential gradient. Hence it is, in general, necessary to use a numerical iteration scheme to get an exact solution where the potential distribution is initially guessed and then improved by iteration. Iterative schemes for obtaining an exact solution for the spherically symmetric sheath about a nonemitting body have been developed by Laframboise[8] and by Parker.[11] The problem of an emitting sphere was discussed by Chang and Bienkowski,[12] who used the analysis scheme of Bernstein and Rabinowitz[13] to obtain solutions for the thin-sheath case, where the Debye length for the emitted electrons was small compared to the radius of the sphere.

We are concerned here with the thick-sheath case, where the Debye length for the emitted electrons is on the order of or larger than the spacecraft dimensions. In order to obtain a tractable problem, we assume spherical symmetry. It has been shown that a spherically symmetric sheath can, at times, be a good approximation in the magnetosphere (see the discussion of the use of a Debye potential in Ref. 14). For the ATS-6 spacecraft, most of the secondary electrons probably come from the large 10-m-diam parabolic reflector, which should provide a reasonably homogeneous source. The reflector is not solid but is of a mesh construction such that it is partially transparent to both sunlight and particles. Hence a spherically symmetric treatment should give a reasonable first approximation to the sheath. This should be especially true during an eclipse, when only secondary electrons are emitted, since the primary particles are incident from all directions. A better approximation would require either particle trajectory calculations[15] or a numerical simulation or modeling as performed by Soop[16] or Cauffman and Maynard.[17]

Because of the spherical symmetry, the equations can be expressed in terms of one spatial variable (r) and the radial and transverse components of velocity (v_r and v_T). It is convenient to transform from velocity to the total energy (E) and angular momentum (J), since the latter are constants of the motion for any particular trajectory. If r is expressed in terms of the probe radius R, and if velocities and energy are expressed in terms of the particle thermal velocity and energy, respectively, then we have the following dimensionless quantities (where e and m are the particle charge and mass, k is Boltzmann's constant, and T the temperature):

$$x = r/R \tag{1}$$

$$E = (1/kT)\left(\frac{1}{2}mv_r^2 + \frac{1}{2}mv_T^2\right) + [e\phi(r)/kT] \tag{2}$$

$$J = rv_T/R\sqrt{2kT/m} \tag{3}$$

We also define the dimensionless potential u(x) and equivalent potential for radial motion U(x) as follows:

$$u = \phi e/kT < 0 \tag{4}$$

for an attracted particle (i.e., $e < 0$ for electrons; $e > 0$ for ions), and

$$U = u + (J^2/x^2) \tag{5}$$

The equivalent potential U(x) determines the turning points of the particle trajectory (see, e.g., Ref. 18). For a given energy E and angular momentum J, motion of particles is possible only for $E > U$, and a turning point in the radial motion (where $v_r = 0$) occurs where $E = U$. This is illustrated in Fig. 1, where we have assumed that the probe is repulsive and that there is an electrostatic potential barrier at some radius denoted by x_B. In general, a barrier in the equivalent potential curve is caused by a combination of electrostatic and angular momentum effects. We shall treat repelled particles now (i.e., electrons) and then attracted particles (ions).

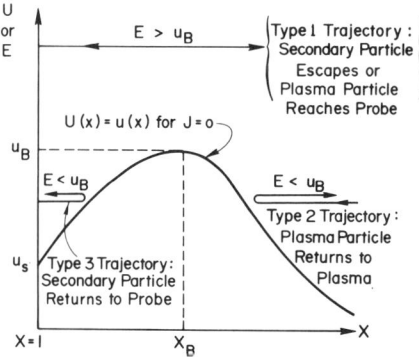

Fig. 1 Equivalent potential curve for zero angular momentum (electrons). E and U are the normalized energy and equivalent potential, and x is the normalized radius.

Electrons

Three types of trajectories are illustrated in Fig. 1. At an energy that is less than the barrier height u_B, particles emitted by the probe with zero angular momentum cannot escape to the plasma and instead return to the probe. Similarly, particles with the same energy in the ambient plasma cannot reach the probe and are reflected back by the potential barrier. At an energy that is greater than u_B, particles emitted by the probe can escape to the plasma, and similarly particles from the plasma can reach the probe. On each of these trajectories, the distribution function is constant and is evaluated by going to the source. For example, a Maxwellian distribution for the ambient plasma would lead to a distribution function proportional to $\exp(-E)$, since u approaches zero as r becomes large. For particles emitted at the probe surface with a Maxwellian distribution, the distribution at any (accessible) radius would be proportional to $\exp(u_S - E)$, where u_S is the normalized surface potential.

Figure 2 illustrates the behavior of the equivalent potential function with increasing angular momentum. As J increases, U increases more rapidly for small values of the radius. Consequently, it is possible for a potential well to

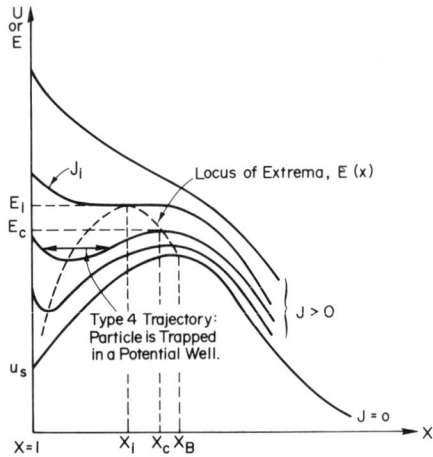

Fig. 2 Equivalent potential curves for different angular momenta (electrons). x_B is the position of the barrier, x_c is the maximum radius for trapped particles, and x_i is the position of the inflection point.

develop where trapped particles can exist, although a collision of some sort is required to get a particle into such a closed orbit. As the angular momentum increases further, the potential well will disappear at a critical angular momentum J_i, where an inflection point occurs at some radius x_i (and energy E_i). For larger values of J, the equivalent potential decreases monotonically outwards from the probe. For every $J < J_i$, there exists a maximum in the equivalent potential function at some radius outside x_i. In general, there also will be a minimum in the curve at some radius inside x_i. (The minimum may not occur always, but this will not affect our results.) The dashed curve in Fig. 2 shows schematically the locus of extrema in the equivalent potential curves. This locus is labeled E(x) (with J as parameter). Since the potential is not known as a function of x, the locus of extrema cannot be obtained in the usual way by differentiation. The locus of maxima marks the boundary between trajectories that can reach the plasma from the probe (and vice versa) and trajectories that return to their source. Hence an accurate determination of the particle density requires a knowledge of E(x).

The particle density at any point x is given by an integral over velocities. This can be transformed into an integral over E and J^2 through the definitions in Eqs. (2) and (3). The results for density n and also for the flux F are, assuming Maxwellian distributions for both plasma and emitted electrons,[19]

$$\frac{n}{n_o} = \frac{e^{u_s}}{2\pi^{1/2} x^2} \iint \frac{we^{-E} \, dE \, dJ^2}{\sqrt{E - u - J^2/x^2}} \quad \text{for secondary particles} \quad (6)$$

$$\frac{n}{n_o} = \frac{1}{2\pi^{1/2} x^2} \iint \frac{we^{-E} \, dE \, dJ^2}{\sqrt{E - u - J^2/x^2}} \quad \text{for plasma particles} \quad (7)$$

$$\frac{F}{F_o} = \frac{e^{u_s}}{x^2} \iint e^{-E} \, dE \, dJ^2 \quad \text{for secondary particles} \quad (8)$$

$$\frac{F}{F_o} = \frac{1}{x^2} \iint e^{-E} \, dE \, dJ^2 \quad \text{for plasma particles} \quad (9)$$

where $E \geq u_s$ for secondary particles and $E \geq 0$ for plasma particles.

In these expressions, n_o is the density of the particles at the source for an unperturbed distribution; the quantity $F_o = n_o (kT/2m)^{1/2}$; the weight w has the value unity in regions where only one-way trajectories are possible (type 1 trajectory) and has the value 2 in regions where particles from a given source can be going in both directions (type 2 and 3 trajectories). The flux has the weight unity, since we are calculating only the one-way flux of particles.

The limits on the integrals in the preceding expressions are determined by the boundaries in the J^2 - E plane which separate the various types of trajectories. Note that the topology of these regions in the J^2 - E plane is independent of the particle temperature. Hence the boundaries can be determined from the normalized quantities, E, J, u, etc. At any radius x, the energy E must be greater than U(x) for a given J:

$$E > u(x) + J^2/x^2 \equiv U(x)$$

or

$$J^2 < x^2 [E - u(x)]$$

The curve that is obtained by replacing the inequality sign by equality is a straight line with slope x^2 in the J^2 - E plane, as shown in Fig. 3. The intercept of this line on the horizontal axis occurs at $E = u(x)$. Trajectories can exist only below this line, which we shall designate the "turning point" line. The dashed line shows the locus of extrema E(x), except that here it is shown as $J^2(E)$ with x as a parameter. The locus has a cusp at (J_i^2, E_i) corresponding to the inflection point in the equivalent potential curve. The branch to the left of the cusp is the locus for minima in the equivalent potential curve, and the branch to the right is the locus for maxima. Each turning point line for $x < x_B$ (x_B is the barrier position) is tangent to the locus curve at the point (J^2, E) corresponding to the location of the extremum at that radius. Consequently, the slope of the locus curve increases monotonically as one progresses along it from left to right; it has a maximum slope (of x_B^2) where it crosses the energy axis at u_B.

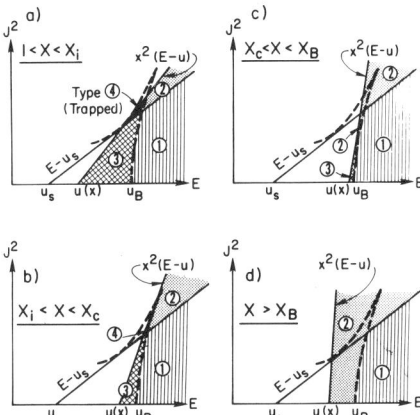

Fig. 3 Regions for different types of trajectories in the J^2 - E plane. Each plot is for a fixed radius (x) (electrons). See Figs. 1 and 2 to identify the four types of trajectories.

Figure 3 distinguishes four possible situations, that is, four different topologies in the J^2 - E plane. Each plot is for a different value of the radius x and shows the types of trajectories which can occur in the different regions. In addition to the special values of the parameters corresponding to the inflection point (J_i, E_i, x_i), it also is necessary to identify the critical values J_c, E_c, which correspond to the case where the value of the equivalent potential at the probe surface equals its value at the maximum. This case also is illustrated in Fig. 2. The radius at which this maximum occurs, x_c, is the largest radius where a trapped particle can exist.

Up to now, the treatment has been exact; i.e., if we knew the locus of extrema curve and if we knew how to calculate the number of particles in trapped orbits, we could obtain an exact solution to the problem. We make the following two approximations: we assume that there are no particles in trapped orbits, and we approximate the locus of maxima curve by a vertical line at u_B. The justification for neglecting trapped particles is that their existence depends upon the presence of some collision mechanism that is unknown but undoubtedly very small for the kinds of low-density plasmas that we are interested in. Also, the time it would take to populate these closed orbits is on the order of several collision times, which is probably long compared to the time scale over which the local plasma can change its characteristics significantly.

The approximation of the locus of maxima curve by the vertical line $E = u_B$ is justified if the sheath is thick. This approximate locus has the right intercept but an infinitely steep slope compared to the correct slope of x_B^2 at that point. However, the true slope of the locus of maxima is always greater than x_i^2 (its value at the cusp), and, if the sheath is thick (i.e., large x_i and x_B), this should be a reasonable approximation. The effect on the density of emitted particles will not be very important, since it mainly affects the upper limit of the energy integral for those particles that are returning to the probe. The approximation will tend to underestimate the density of these particles. Outside the barrier position x_B, the locus is taken to be the turning point line for x_B.

With this approximation, there are only three different topologies that need to be distinguished, as shown in Fig. 4. All of the integrals may be obtained analytically, with the

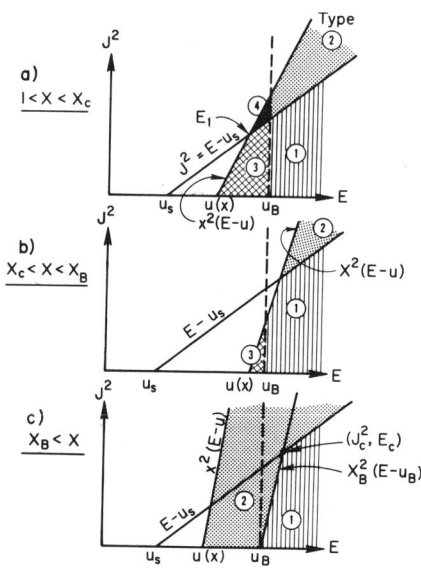

Fig. 4 Regions in the $J^2 - E$ plane for the thick-sheath approximation (electrons). See Figs. 1 and 2 to identify the four types of trajectories.

following results. For photoelectrons or secondary electrons, we have the following:

$\underline{1 < x < x_c}$

$$\frac{n}{n_o} = e^{u_s} \left\{ \frac{e^{-u}}{2} [1 + \mathrm{erf}(u_B - u)^{1/2}] \right.$$

$$- \frac{e^{-E_1}}{2} \left(1 - \frac{1}{x^2}\right)^{1/2} [1 + \mathrm{erf}(u_B - E_1)^{1/2}]$$

$$\left. + \frac{e^{-u_B}}{\pi^{1/2}} \left[\left((u - \frac{1}{x^2})(u_B - E_1)\right)^{1/2} - (u_B - u)^{1/2} \right] \right\} \qquad (10)$$

$\underline{x_c \le x \le x_B}$

$$\frac{n}{n_o} = e^{u_s} \left\{ \frac{e^{-u}}{2} [\mathrm{erf}(u_B - u)^{1/2} + 1] - \frac{1}{2}\left(1 - \frac{1}{x^2}\right)^{1/2} e^{-E_1} \right.$$

$$\left. - \left(\frac{u_B - u}{\pi^{1/2}}\right)^{1/2} e^{-u_B} \right\} \qquad (11)$$

$\underline{x_B \le x}$

$$\frac{n}{n_o} = e^{u_s} \left\{ \frac{e^{-u}}{2} [1 - \mathrm{erf}(u_B - u)^{1/2}] \right.$$

$$- \frac{1}{2}\left(1 - \frac{x_B^2}{x^2}\right) e^{E_2} [\mathrm{erf}(E_c + E_2)^{1/2} - \mathrm{erf}(u_B + E_2)^{1/2}]$$

$$- \frac{1}{2}\left(1 - \frac{1}{x^2}\right)^{1/2} e^{-E_1} [1 - \mathrm{erf}(E_c - E_1)^{1/2}] + \frac{(u_B - u)^{1/2}}{\pi^{1/2}} e^{-u_B}$$

$$+ \frac{(1 - x_B^2/x^2)^{1/2}}{\pi^{1/2}} \left[(E_c + E_2)^{1/2} e^{-E_c} - (u_B + E_2)^{1/2} e^{-u_B} \right]$$

$$\left. - \frac{(1 - 1/x^2)^{1/2}}{\pi^{1/2}} (E_c - E_1)^{1/2} e^{-E_c} \right\} \qquad (12)$$

where

$$E_1 = (x^2 u - u_s)/(x^2 - 1) \tag{13}$$

$$E_c = (x_B^2 u_B - u_s)/(x_B^2 - 1) \tag{14}$$

$$E_2 = (x_B^2 u_B - x^2 u)/(x^2 - x_B^2) \tag{15}$$

For plasma electrons, we have the following:

$\underline{1 < x < x_c}$

$$\frac{n}{n_o} = \frac{e^{-u}}{2} [1 - \mathrm{erf}(u_B - u)^{1/2}] + \frac{(u_B - u)^{1/2}}{\pi^{1/2}} e^{-u_B}$$

$$+ \left(1 - \frac{1}{x^2}\right)^{1/2} \left[\frac{(u_B - E_1)^{1/2}}{\pi^{1/2}} e^{-u_B} \right.$$

$$\left. + \frac{e^{-E_1}}{2} \left(1 - \mathrm{erf}(u_B - E_1)^{1/2}\right) \right] \tag{16}$$

$\underline{x_c < x < x_B}$

$$\frac{n}{n_o} = \frac{e^{-u}}{2} [1 - \mathrm{erf}(u_B - u)^{1/2}] + \frac{1}{2}\left(1 - \frac{1}{x^2}\right) e^{-E_1}$$

$$+ \frac{(u_B - u)^{1/2}}{\pi^{1/2}} e^{-u_B} \tag{17}$$

$x_B \leq x$

$$\frac{n}{n_o} = \frac{e^{-u}}{2}[1 + \mathrm{erf}(u_B - u)^{1/2}] - \frac{(u_B - u)^{1/2}}{\pi^{1/2}} e^{-u_B}$$

$$+ \frac{[1 - (1/x^2)]^{1/2}}{\pi^{1/2}} (E_c - E_1)^{1/2} e^{-E_c}$$

$$+ \frac{1}{2}\left(1 - \frac{1}{x^2}\right)^{1/2} e^{-E_1} [1 - \mathrm{erf}(E_c - E_1)^{1/2}]$$

$$+ \frac{(1 - x_B^2/x^2)^{1/2}}{\pi^{1/2}} \left[(u_B + E_2)^{1/2} e^{-u_B} - (E_c + E_2)^{1/2} e^{-E_c}\right]$$

$$+ \frac{1}{2}\left(1 - \frac{x_B^2}{x^2}\right) e^{E_2} \left[\mathrm{erf}(E_c + E_2)^{1/2} - \mathrm{erf}(u_B + E_2)^{1/2}\right] \quad (18)$$

In these equations, x_c (the maximum radius for trapped particles) is obtained from the fact that it is a root of the equation $E_1(x) = u_B$. Note that it is not necessary to know the barrier position x_B until one has reached the region outside the barrier where $x > x_B$. Also note that one must be careful to use the appropriate temperature in an actual calculation, since all of the energies and potentials have been normalized to the temperature of the particles that are being considered.

Ions

The analysis for ions is very similar to that for the electrons. For ions, $u(x)$ is negative, and the equivalent potential curves have the shapes shown in Fig. 5. What was a potential barrier for the electrons at x_B is a potential well for the ions. However, for $J > 0$, a potential barrier develops for the ions at radii outside x_B. This barrier is caused by angular momentum effects only and is present whenever the potential $u(x)$ falls off more rapidly than x^{-2}. The locus of extrema curve is shown as the dashed line in Fig. 5. It has a maximum at the inflection point which separates minima from maxima, as for the electron case. However, the locus for maxima in this case approaches the energy axis ($E = 0$) at $x = \infty$.

The different topologies that can occur in the J^2-E plane are shown in Fig. 6. The locus of extrema curve approaches the energy axis at $E = 0$ with an infinite slope.

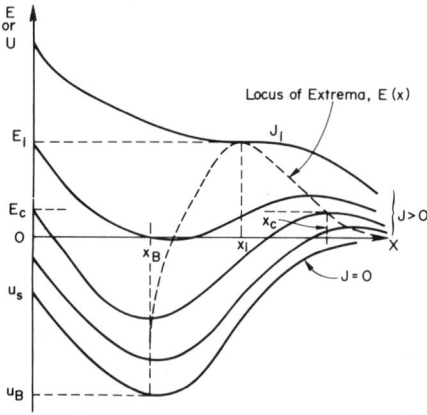

Fig. 5 Equivalent potential curves for ions. x_B is the position of the barrier, x_i is the position of the inflection point, and x_c is the maximum radius for trapped particles.

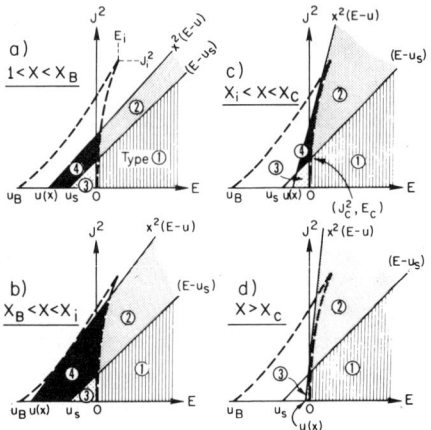

Fig. 6 Regions for different types of trajectories in the J^2-E plane (ions). See Figs. 1 and 2 to identify the four types of trajectories.

Trapped particles on closed orbits can occur in the regions marked as "type 4 trajectories." Regions where type 3 trajectories can occur corresponding to orbits intersecting the probe at both ends of the trajectory will be assumed to be unpopulated.

We again make the assumption that there are no trapped particles. In order to obtain analytic expressions for the ion density, we shall use two different approaches. The first approach is to assume, as for the electron case, that the locus of maxima curve can be approximated as a vertical line at $E = 0$. This has both the right intercept at $E = 0$ and also the right slope at that point. With this assumption, only two cases need to be distinguished, and analytic expressions may be obtained for the ion density with an assumed Maxwellian distribution. These expressions are given below, with the two different topologies shown in Fig. 7.

However, it is difficult to estimate the accuracy of this approximation even though the assumed locus has the right slope and intercept at $E = 0$. This is because the assumption

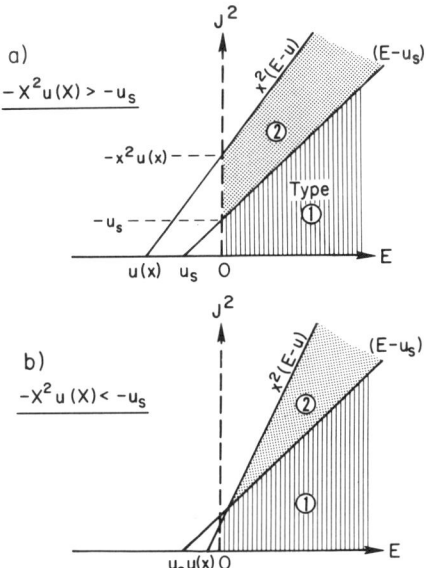

Fig. 7 Regions in the $J^2 - E$ plane for the thick-sheath approximation (ions). See Figs. 1 and 2 to identify the four types of trajectories.

affects primarily the lower limit of the energy integral where the distribution function has the largest value. Consequently, we use a second approach to get another estimate for the ion density. We assume a monoenergetic ion distribution at the energy E_o. If E_o is greater than E_i, the density does not depend upon the position or shape of the locus of extrema curve. This expression also is given below. Note that, for both approaches, the errors in the approximations can be evaluated after the fact (i.e., after the potential distribution has been calculated) by numerically obtaining the extremum curve $E(x)$:

Maxwellian Distribution

$-x^2 u > -u_s$

$$\frac{n}{n_o} = \left(\frac{-u}{\pi}\right)^{1/2} + \frac{e^{-u}}{2}[1 - \mathrm{erf}(-u)^{1/2}] + \frac{1}{x}\left(\frac{u_s - x^2 u}{\pi}\right)^{1/2}$$

$$+ \frac{1}{2}\left(1 - \frac{1}{x^2}\right)^{1/2} \exp\left\{\frac{u_s - x^2 u}{x^2 - 1}\right\}\left[1 - \mathrm{erf}\left(\frac{u_s - x^2 u}{x^2 - 1}\right)^{1/2}\right] \quad (19)$$

$-x^2 u < -u_s$

$$\frac{n}{n_o} = \left(\frac{-u}{\pi}\right)^{1/2} + \frac{e^{-u}}{2}[1 - \mathrm{erf}(-u)^{1/2}]$$

$$+ \frac{1}{2}\left(1 - \frac{1}{x^2}\right)^{1/2} \exp\left\{\frac{u_s - x^2 u}{x^2 - 1}\right\} \quad (20)$$

Monoenergetic Distribution

$$\frac{n}{n_o} = \frac{1}{2}\left[(1-u)^{1/2} + \left(1 - u - \frac{1}{x^2} + \frac{u_s}{x^2}\right)^{1/2}\right] \text{ for } E_o > E_i \quad (21)$$

For the monoenergetic distribution, the potential and energies have been normalized to the ion energy E_o. If we are interested only in the region $1 < x < x_B$, then expression (21) is valid for $E_o > E_B$ ($E_B < E_i$), where E_B is defined by the intersection of the turning point line for x_B with the locus of maxima curve.

Poisson's Equation

Let the subscripts 1 through 3 refer, respectively, to photoelectrons, plasma electrons, and ions. Then Poisson's equation may be written as follows:

$$\frac{d^2 u_1}{dx^2} + \frac{2}{x}\frac{du_1}{dx} = \frac{-1}{L_1^2}\left[\frac{n_1(x)}{n_{10}} + \frac{n_{20}}{n_{10}}\frac{n_2(x)}{n_{20}} - \frac{n_{20}}{n_{10}}\frac{n_3(x)}{n_{20}}\right] \quad (22)$$

where L_1 is the ratio of the photoelectron Debye length to the probe radius, and where the ambient plasma is assumed to be neutral. Note again that in evaluating the densities we must be careful to use the appropriate temperatures for each of the separate space charge terms. Note also that the dimensionless potential u_1 has the opposite sign of the true potential.

By transforming to a new variable, $y = u_1 x$, the first derivative may be eliminated to obtain the equation in the form

$$\frac{d^2 y}{dx^2} = \frac{-x}{L_1^2}\left[\frac{n_1(x)}{n_{10}} + \frac{n_{20}}{n_{10}}\frac{n_2(x)}{n_{20}} - \frac{n_{30}}{n_{10}}\frac{n_3(x)}{n_{20}}\right] \quad (23)$$

This may be integrated numerically from the probe surface (x = 1) outwards. The potential is known at the surface, and the barrier height and particle densities and temperatures are assumed to be known. One boundary condition that is not known is the value of the potential gradient at the probe surface. Hence, what must be done in practice is to try various values for the starting slope until the correct one is found. The correct one is identified by the fact that it leads to a barrier height u_B of the correct value.

Results

Equation (23) was integrated numerically using a fourth-order Runge-Kutta technique. The problem was programmed on an interactive computer system so that the starting slope could be changed by the operator. Input data for the program were the ratios of the ambient particle densities and temperatures to the photoelectron density and temperature, the normalized satellite potential and barrier height, and the photoelectron Debye length in units of the satellite radius (assumed to be 5 m). The program also asks for the increment in radius and for the starting slope. The accuracy of the integration was tested by choosing smaller increments until there was no significant change in the results. The results obtained so far have been with the assumption of a monoenergetic distribution for the ions.

If a potential barrier exists outside the probe, the normalized potential ($u_1 > 0$) should start out at the surface with a positive slope and a negative curvature. The slope should decrease and go through zero at the barrier position, and then at some point outside the barrier position the curvature should change sign near the point where the space charge goes through zero. [From (23) it can be seen that it is actually the curvature of y that vanishes with the space charge.] In practice, what happens if too large a starting slope is chosen is that the space charge goes through zero, and hence the curvature changes from negative to positive, whereas the slope is still positive. As a result, the potential will keep on increasing indefinitely. If a lower starting slope is chosen, then it is possible for a barrier to develop. The largest possible value for the height of the potential barrier occurs when both the slope and curvature go through zero simultaneously. (Strictly speaking, this is not a physically realistic solution, since the potential in this case never would approach zero outside the barrier, but this is the limiting case for the existence of a barrier.)

The program was applied to the four sets of ATS-6 data shown in Table 1 (see Ref. 10 for more details). The table

Table 1 ATS-6 data used in study

	Day 198 7-17-74	Day 199 7-18-74	Day 204 7-23-74	Day 273 9-30-74
Secondary electron density, cm^{-3}	10.3	90	200	0.4
Plasma density, cm^{-3}	0.2	1.2	90	17
Photoelectron temperature, eV	4.9	2	1.9	6
Plasma electron temperature, eV	65	32	14.5	320
Plasma ion temperature, eV	7	10	3.5	650
Photoelectron Debye length, m	5.2	1.1	0.73	29.4
Satellite potential, V	-20	0	-2	-2000
Transition energy, $\phi_B - \phi_s$, V	-40	-10	-8	-50
Barrier potential, ϕ_B, V	-60	-10	-10	-2050
Dimensionless barrier potential, $u_B = \phi_B e/kT_1$	12.2	5	5.3	342
Maximum calculated value for u_B	5.0	3.1	1.7	None
Position of calculated barrier, $x_B = r_B/R$	6	4	1.5	...

shows the particle properties and the observed satellite and barrier potentials. In each of the four cases, it was impossible to reproduce the observed barrier height. The maximum barrier potential that could be obtained is shown in the table. In every case, it is significantly lower than the observed value. The reason for this behavior is that at the observed barrier potential the ion density always was enhanced over its ambient value, whereas the plasma electron density always was smaller than the ambient value. The difference between these two densities always was larger than the density contributed by the secondary electrons, so that it was impossible to maintain a negative space charge.

The plasma electron density that was used in the calculation was that for energies below about 100 eV (except for day 273, where the listed electron density describes the distribution out to about 500 eV). The addition of a higher-energy component of plasma electrons has the effect of reducing the possibility of the existence of a potential barrier, because, in order to maintain neutrality in the ambient plasma, the ion density (assumed to be equal to the electron density) must be increased also. Actually, the observed ion densities (not shown) tend to be somewhat higher than the electron densities if only the low-energy electron component is counted, as in the table. However, the higher the ambient particle density, the more difficult it is for a barrier to develop. The problem is to get rid of the ions. For example, if equal numbers of high-energy ions and electrons were assumed to exist in the ambient plasma which have not been included in the calculation, including them would not make the development of a barrier any easier, since the space charge already is too positive in the sheath.

Probably the most convincing case is the eclipse data of day 273. Here it was impossible to develop a potential barrier for any value of the starting slope. The large secondary electron Debye length should make the thick sheath approximation very good. The space charge at the surface was positive due to the large attractive potential for the ions and remained positive at larger radii.

We conclude that the barrier potentials inferred from the ATS-6 data are too large to be explained in terms of a spherically symmetric photoelectron or secondary sheath surrounding a uniformly charged spacecraft. Taking into account the actual geometry of the spacecraft probably will not change this conclusion. The most probable explanation is that different portions of the spacecraft surface are charged to different potentials.

References

[1] Whipple, E. C., Jr. and Parker, L. W., "Effects of Secondary Electron Emission on Electron Trap Measurements in the Magnetosphere and Solar Wind," Journal of Geophysical Research, Vol. 74, November 1969, p. 5763.

[2] Singer, S. F. and Walker, E. H., "Photoelectric Screening of Bodies in Interplanetary Space," Icarus, Vol. 1, January 1962, p. 7.

[3] Medved, D. B., "On the Formation of Satellite Electron Sheaths Resulting from Secondary Emission and Photoeffects," Interactions of Space Vehicles with an Ionized Atmosphere, edited by S. F. Singer, International Series of Monographs in Aeronautics and Astronautics, Vol. 18, Pergamon Press, New York, 1965.

[4] Grard, R. J. L. and Tunaley, J. K. E., "Photoelectron Sheath near a Planar Probe in Interplanetary Space," Journal of Geophysical Research, Vol. 76, April 1971, p. 2498.

[5] Guernsey, R. L. and Fu, J. H. M., "Potential Distribution Surrounding a Photo-Emitting Diode in a Dilute Plasma," Journal of Geophysical Research, Vol. 75, June 1970, p. 3193.

[6] Fu, J. H. M., "Surface Potential of a Photoemitting Plate," Journal of Geophysical Research, Vol. 76, April 1971, p. 2506.

[7] Schroder, H., "Spherically Symmetric Model of the Photoelectron Sheath for Moderately Large Plasma Debye Lengths," Photon and Particle Interactions with Surfaces in Space, edited by R. J. L. Grard, Reidel Publishing Co., Dordrecht-Holland, 1973, p. 51.

[8] Laframboise, J. G., "Theory of Spherical and Cylindrical Langmuir Probes in a Collisionless Maxwellian Plasma at Rest," UTIAS Rept. 100, 1966, University of Toronto.

[9] Tunaley, J. K. E. and Jones, J., "The Photoelectron Sheath around a Spherical Body," Photon and Particle Interactions with Surfaces in Space, edited by R. J. L. Grard, Reidel Publishing Co., Dordrecht-Holland, 1973, p. 59.

[10] Whipple, E. C., Jr., "Observation of Photoelectrons and Secondary Electrons Reflected from a Potential Barrier in the Vicinity of ATS 6," Journal of Geophysical Research, Vol. 81, February 1976, p. 715.

[11] Parker, L. W., "Computer Solutions in Electrostatic Probe Theory, I. Spherical Symmetry with Collisions," TR AFAL-TR 72-222, Pt. I, 1973, Air Force Avionics Lab., Wright-Patterson Air Force Base, Ohio.

[12] Chang, K. W. and Bienkowski, G. K., "Effects of Electron Emission on Electrostatic Probes at Arbitrary Pressures," The Physics of Fluids, Vol. 13, April 1970, p. 902.

[13] Bernstein, I. B. and Rabinowitz, I. N., "Theory of Electrostatic Probes in a Low-Density Plasma," The Physics of Fluids, Vol. 2, March-April 1959, p. 112.

[14] Whipple, E. C., Jr., Warnock, J. M., and Winkler, R. H., "Effects of Satellite Potential on Direct Ion Density Measurements through the Plasmapause," Journal of Geophysical Research, Vol. 79, January 1974, p. 179.

[15] Parker, L. W. and Whipple, E. C., Jr., "Theory of Spacecraft Sheath Structure, Potential, and Velocity Effects on Ion Measurements by Traps and Mass Spectrometers," Journal of Geophysical Research, Vol. 75, September 1970, p. 4720.

[16] Soop, M., "Report on Photosheath Calculations for the Satellite Geos," Planetary and Space Science, Vol. 20, June 1972, p. 859.

[17] Cauffman, D. P. and Maynard, N. C., "A Model of the Effect of the Satellite Photosheath on a Double Floating Probe System," Journal of Geophysical Research, Vol. 79, June 1974, p. 2427.

[18] Goldstein, H., Classical Mechanics, Addison-Wesley, Cambridge, Mass., 1953.

[19] Grard, R. J. L., "Properties of the Satellite Photoelectron Sheath Derived from Photoemission Laboratory Measurements," Journal of Geophysical Research, Vol. 78, June 1973, p. 2885.

SPACECRAFT POTENTIAL CONTROL WITH ELECTRON EMITTERS

R. Grard,* A. Gonfalone,* and A. Pedersen*
European Space Agency, Noordwijk, The Netherlands

Abstract

Negative charging of spacecraft occurs in magnetospheric environments whenever the outgoing flux of photoelectrons cannot balance the incoming flux of ambient electrons. Large negative surface potentials often are observed in energetic plasmas, and levels of the order of several kilovolts have been measured in the outer-Earth magnetosphere. This situation certainly also will be met in the vicinity of Jupiter, where the photoemission rate is 30 times less than at Earth orbit. Such surface phenomena are a source of interference for scientific experiments and are responsible for anomalies in the behavior of subsystems. It is possible to control the potential of a conductive body by releasing the accumulated negative charge from an electron-emitting probe, which can supplement or even replace photoemission. Electrons can be extracted from a metal by thermoionic emission or by field emission. The merits of the two systems are evaluated and compared: an electron field emitter requires no power supply; however, an electron gun fitted with a heated cathode allows a finer adjustment of the spacecraft potential. Both types of electron emitters also can be used as stable references, with respect to which the spacecraft potential can be biased, even at positive potentials. Information about ambient density and temperature can be obtained at the cost of little additional hardware.

Introduction

A spacecraft immersed in a plasma acquires an equilibrium potential such that the sum of the currents collected by its

---

Presented as Paper SM-127F at the Spring Annual Meeting of the American Geophysical Union, Washington, D.C., June 16-19, 1975.

*Member of Professional Staff, Space Sciences Department.

surface is zero. In other words, the difference between the incoming ambient electron and ion flows is balanced exactly by the rate at which electrons are released by secondary and photoemission. In a tenuous and relatively cold plasma, such as the solar wind, photoemission is preponderant generally for objects irradiated partly by sunlight. Under these conditions, a body develops a positive potential that controls the escape of the photoemitted electrons; this potential, of the order of the electron mean kinetic potential, is equal to a few volts. (The kinetic potential of an electron is given by the magnitude of the accelerating voltage associated with its kinetic energy; the kinetic potential, in volts, therefore is measured by the same number as the kinetic energy, in electron volts).

In a dense plasma, like the lower ionosphere, or in highly energetic environment, such as the outer magnetospheres of Earth and other planets, the saturation current of the ambient plasma electrons may predominate over that of other species; this is true, in particular, for bodies in shadow or in eclipse. The surface reaches a potential sufficiently negative to limit the current of ambient electrons and make it equal to the combined contribution of the ambient ions, secondary electrons, and photoelectrons. In this situation, the body potential is equal to a few times the plasma electron mean kinetic potential, and it can be overcome only by electrons having energies several times larger than the thermal energy. The electron mean kinetic potential is typically 1 kv in the outer magnetosphere of the Earth, and it is not uncommon that spacecraft on geostationary orbit charge up to negative potential [1] of several kilovolts. This phenomenon is a source of interference for scientific experiments because it disturbs the density and energy distribution of the ambient-charged particles.

This work describes a system that can prevent a spacecraft, covered with conductive coating and paint, from reaching highly negative voltages in energetic plasmas. It is demonstrated that an electron emitter can release the negative charge accumulated on a conductive surface and reduce considerably the floating potential of the spacecraft. The emission of electrons from a metal in a vacuum is a function of its temperature and of the electric field existing at its surface.[2] Thermoionic emission corresponds to the case wherein the temperature of the electrode (cathode) is the predominant factor; field emission describes a phenomenon occuring at ambient temperature when large electric fields exist at the tip of a negatively charged point. This work describes the two types of emitters and compares their merits.

It also is shown that the potential of an electron emitter is very stable and that the whole spacecraft can be biased with respect to this reference electrode if a voltage source is available.[3]

Spacecraft Environment

Typical plasma environments are described in Table 1, where: n = the plasma density, ϕ_e = the plasma electron kinetic potential, λ_D = the debye length, j_e = the plasma electron saturation current density, j_{ph} = the photoelectron saturation current density, ϕ_{f0} = the floating potential of a surface in shadow, and ϕ_{f1} = the floating potential of a spherical conductive object in sunlight. Secondary electron emission is not taken into account. The plasma is assumed to be made of electrons and protons in thermal equilibrium, and the floating potential in absence of photoemission, [6] ϕ_{f0} is of the order of 3.8 ϕ_e. In sunlight, [7] the floating potential is taken to be $\phi_{f1} = \phi_{ph} \ln(j_{ph}/4j_e)$ when $j_e < j_{ph}/4$ and $\phi_{f1} = \phi_e \ln(4j_e/j_{ph})$ when $j_e > j_{ph}/4$; the factor 4 stands for the fact that the collecting area of a spherical object is four times larger than its photoemitting area, and the photoelectron mean kinetic potential, [8] ϕ_{ph} is of the order of 1.5 volt. It must be noted that the floating potential of a conductive body is probably always negative in the magnetospheres of Jupiter and Uranus, even in sunlight. In Earth environment, this potential generally is positive in sunlight, because the current density of photoelectrons is larger than that of ambient electrons most of the time.

Table 1 Surface potentials in space environments

Parameter	Magnetosphere			Solar wind[4,5]		
	Earth	Jupiter	Uranus	1 AU	5 AU	20 AU
$n(cm^{-3})$	1	1	0.1	7	0.2	0.02
$\phi_e(V)$	1000	1000	100	15	3.2	0.3
$\lambda_D(m)$	235	235	75	10	30	29
$j_e(\mu Am^{-2})$	0.85	0.85	0.027	0.73	0.01	0.001
$\frac{1}{4}j_{ph}(\mu Am^{-2})$	5	0.2	0.0125	5	0.2	0.0125
$\phi_{f0}(V)$	-3800	-3800	-380	-56	-12	-1
$\phi_{f1}(V)$	2.65	-1500	-77	2.9	4.9	7.2

Working Principle of an Electron Emitter

Consider a system made of an electron emitter connected to a spacecraft through a voltage source and an ammeter, as shown in Fig.1. The current characteristic of the emitter and spacecraft is represented qualitatively as a function of potential; the voltage reference is that of the undisturbed plasma at large distances from the spacecraft. The following nomenclature is used:

i_p = electron emitter current
i_{s0} = current collected by the spacecraft in shadow, or when the photoemission rate is insufficient
i_{s1} = current collected by the spacecraft in sunlight, with high photoemission rate
I_{es} = saturation current associated with the collection of ambient electrons by the spacecraft
I_{ps} = saturation current of the emitter
ϕ_{s0} = floating potential of spacecraft (low photoemission rate)
ϕ_{s1} = floating potential of spacecraft (high photoemission rate)

When the emitter is connected to the spacecraft, the current flowing through the system is $i_p = i_s = i_b$, and the probe and spacecraft potentials, ϕ_p and ϕ_s, are linked by the relationship $\phi_p - \phi_s = V_b$, where V_b is the bias voltage. When $V_b = 0$, the floating potential is defined simply by the intersection of the probe characteristic with the spacecraft current curve. In shadow, the new floating potential ϕ_0 is closer considerably

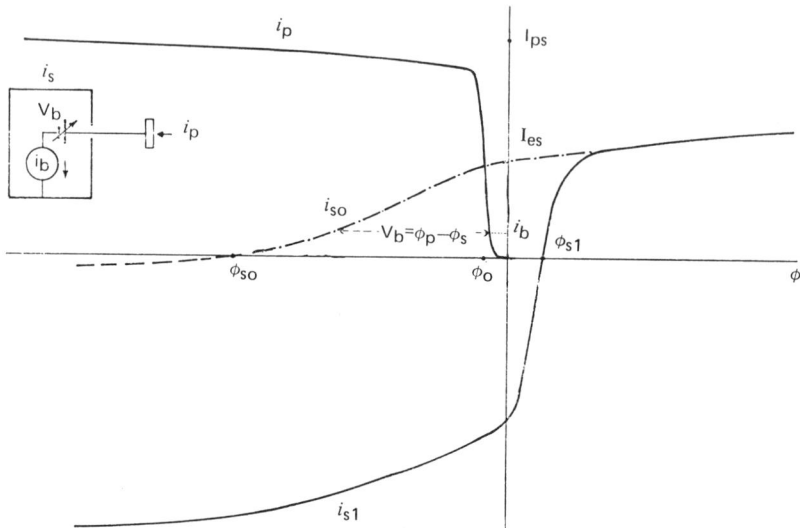

Fig. 1 Current - Voltage characteristics of emitter and spacecraft.

to zero than ϕ_{so}; in sunlight the potential $\phi_1 = \phi_{s1}$ is unchanged.

It will be shown that the emitter current varies rapidly with the applied voltage; this device provides a stable reference with respect to which the spacecraft potential may be biased. The spacecraft current characteristic therefore can be measured in a voltage range, which is limited only by the capability of the bias voltage generator. In other words, the spacecraft itself behaves like a Langmuir probe. The electron saturation current collected by the vehicle, in particular, can be estimated under all circumstances. The main condition to be satisfied, which can be fulfilled easily, is that the emitter's current I_{ps} must be larger than the saturation current I_{es} collected by the spacecraft.

Thermoionic Emitter

Electron emission from a cathode is space-charge limited; the current through a diode is of the form $I = K\, V^{3/2}/\alpha^2$ where K and α^2 are a function of the geometry of the diode.[9] If I is measured in amps and V in volts, $K \simeq 3 \times 10^{-5}$ for spherical symmetry and α is a function of the ratio r_a/r_c of the anode-to-cathode radii. Let us assume first that the emitter is mounted away from the spacecraft at the tip of a boom and that spherical symmetry applies; r_c is given by the physical dimension of the cathode, and r_a is a distance over which space-charge neutrality is restored in the plasma, that is, the debye length λ_D.

The straight lines of Fig. 2 represent the cathode currents against its potential with respect to space (not with respect to the spacecraft) for values of the ratio r_a/r_c ranging from 10 to 10^6. Also shown is the electron current collected by a spherical conductive spacecraft 2 m in radius, in various plasma environments, photoelectron and ion currents are neglected. The value of the potential, for which the current emitted by the cathode is equal to the current collected by the spacecraft, is the floating potential of the system. It is seen that this potential is not influenced strongly by the ratio r_a/r_c and that it will not be less than - 10 v in typical magnetospheric environments. In fact, the clamping potential even may be more positive than actually shown; since the electrons are emitted with a finite energy, the current does not vanish when the potential goes to zero.

An upper limit of the perturbation caused by the presence of the vehicle can be estimated readily. In the preceeding

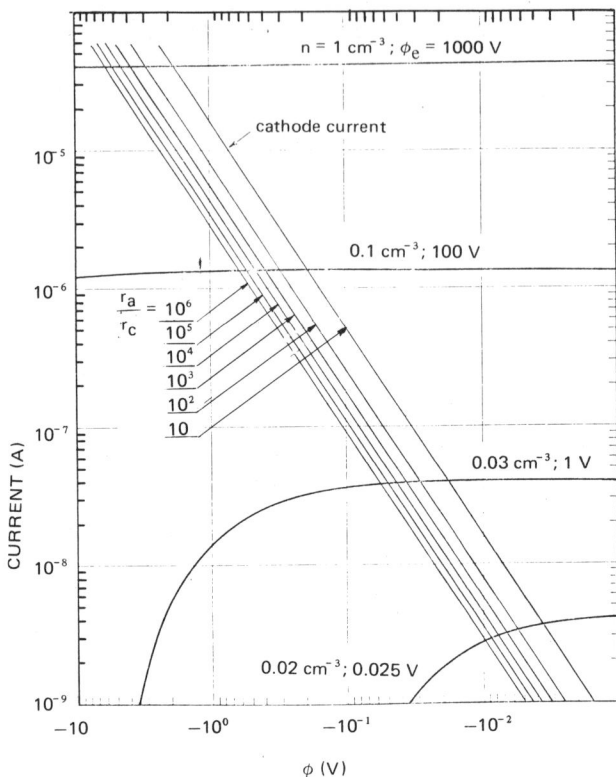

Fig. 2 Determination of the clamping potential with cathode.

approach, the separation between the probe and the spacecraft is assumed to be infinite, and the probe potential is referred to that of the undisturbed plasma ($V = \phi$). In the case of finite probe spacecraft separation, one may take as reference the potential generated by the vehicle in the vicinity of the cathode; then the potential drop which determines the electron current is reduced to approximately $V \simeq \phi(1-r_s/r)$, where r_s is the radius of the body (assumed to be spherical) and $r-r_s$ is the separation between the cathode and the spacecraft surface. The results are that $V \simeq \phi/2$ for a separation equal to r_s; the influence of the satellite body therefore can be taken into account in Fig. 2 by shifting the cathode current characteristics along the horizontal axis by an amount equal to log (V/ϕ). It is seen then that the clamping potential for a separation, equal to r_s is approximately twice the value obtained for infinite separation. For this reason, it is acceptable to position the cathode at a distance of the order of r_s and still obtain efficient potential clamping. Biasing the spacecraft with respect to the emitter can be achieved with a voltage source as shown on Fig. 2. When the cathode is mounted on the surface of the spacecraft, it is anticipated that space-charge-effects will be more important. In this situation, planar geometry may be more representative than the spherical symmetry assumed here. It has been demonstrated with ATS-5 and 6, however, that spacecraft potentials even could be controlled with an emitter mounted in an open cavity under the surface of the vehicle.[10] A grid system, in this case, certainly would help electron extraction; this type of consideration naturally leads to the concept of an electron gun, such as the one proposed for the International Sun Earth Explorer/A mission.[11]

Field Emitter

Electrons are emitted from a cold metal[12] when the electric field on the surface is of the order of 10^9 v m^{-1}. The potentials at work in a magnetospheric environment are of the order of 1 kv; field emission is, therefore, insignificant for bodies of the size of a spacecraft but is important for particles with dimensions less than 1 µm.

Let us consider now a probe made of sharp-pointed filaments, electrically connected to the spacecraft but at a distance equal to a few times the typical dimension of the vehicle, as shown in the insert of Fig. 3. This separation is necessary to insure that the strength of the electric field at the tips of the wires is not reduced by the charge induced in the surface of the main body. It can be demonstrated, in these conditions, that the negative charge accumulated on the conductive elements of the vehicle can be released through the probe by field emission.[3]

It is assumed in the following that the spacecraft is a conductive sphere, 1 m in radius, and that the density and mean kinetic energy of the ambient plasma are n = 1 cm^{-3} and $e\phi_e$ = 1 kev. The separation between the probe, made of 100 tips, and the surface is considered to be much larger than the sphere radius; in fact, for a distance of 3 radii, the surface potential is within 25% of the value obtained for infinite separation. The strength of the electric field at the tip of each emitter is, therefore, of the order of $|\phi|/a$, and the emissive area is taken to be $2\pi a^2$, where a = 0.1 µm is the curvature radius of the tip.

The current characteristics of probe and spacecraft, i_p and i_s, are shown in Fig. 3. The potential of the vehicle, equal to - 3800 v without emitter (3.8 ϕ_e), is increased to - 316 v when connected to the probe. The maximum current that

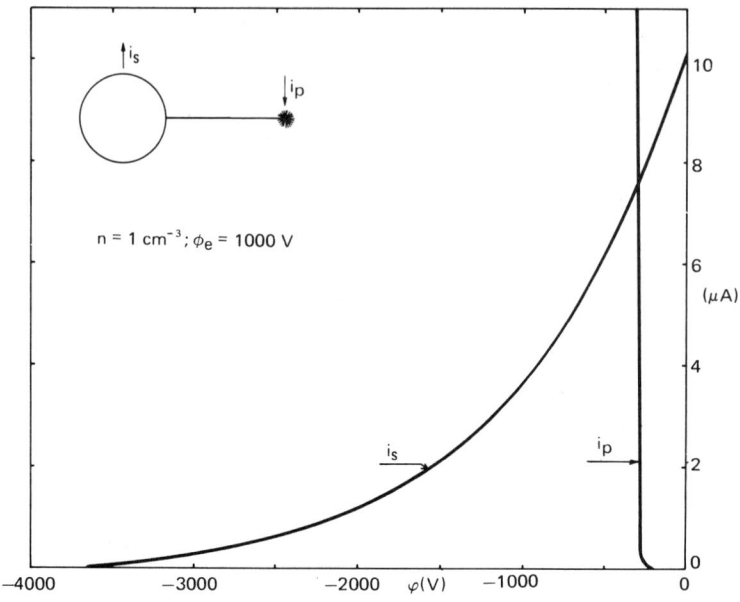

Fig. 3 Clamping potential with field emitter.

can be emitted from such a probe is limited by thermal dissipation, not by space charge; it is of the order of 6 ma for a tungsten probe made of 100 tips with curvature radii 0.1 μm, which is about three orders of magnitude larger than the current actually needed in an environment such as the one considered in Fig. 3.

Table 2. Main features of three potential clamping devices

	Cathode	Gun	Field emitter
Weight (with boom)	150 g	200 g	150 g
Heating power	1 w	1 w	0
Boom length	1 m	0 m	3 m
Clamping potential	−10 V	−10 V	−300 V
Circuitry	heater	heater, grids	none
Main problem	cathode activation	cathode activation	tips damage
Similar previous experiment	ATS-5 and 6	ATS-5 and 6	none

Conclusion

Spacecraft-charging control and plasma diagnostic with electron emitters offer no particular difficulty. Both of the envisaged systems have inherent advantages and limitations. A thermal emitter is more efficient, since it will clamp the spacecraft at potentials above - 10 v, whereas the field emitter capabilities are more limited, in spite of the fact that the emitter requires no heating power. A cathode, contrary to an emission probe, does not have to be mounted on a boom. However, space-charge limitation is more important for a cathode mounted on the spacecraft; the use of grids then may have to be envisaged at the expense of simplicity. The outstanding features of a cathode, an electron gun, and a field emitter are summarized tentatively in Table 2. For reasons of efficiency, weight, simplicity, and reliability, it appears that a simple cathode mounted on a short boom should be the preferred solution for a planetary mission.

References

[1] DeForest, S., "Spacecraft Charging at Synchronous Orbit," Journal of Geophysical Research, Vol. 77, Feb. 1972, pp. 651-659.

[2] Rohrback, F., "Sur les Mécanismes qui conduisent a la formation d'étincelles electrique à très haute tension et sous ultra vide par la mesure des temps de retard à la disruption," European Organization for Nuclear Research, Geneva, Switzerland, Rept. CERN 71-28, 1971.

[3] Grard, R., "Spacecraft Potential Control and Plasma Diagnostic Using Electron Field Emission Probes," Space Science Instrumentation, Vol. 1, Aug. 1975, pp. 363-376.

[4] Dryer, M., Rizzi, A.W., and Shen W.W., "Interaction of the Solar Wind with the Outer Planets," Astrophysics and Space Science, Vol. 22, June 1973, pp. 329-351.

[5] Vogt, R.E., and Siscoe, G.L., "Particles and Fields in the Outer Solar System," Icarus, Vol. 24, March 1975, pp. 333-347.

[6] Self, S.A., "Exact Solution of the Collisionless Plasma Sheath Equation," The Physics of Fluids, Vol. 6, Dec. 1963, pp. 1762-1768.

[7] Stauning, P., "Charged Particle Emission from a Geostationary Satellite," The ESRO Geostationary Magnetospheric Satellite, European Space Research Organization, SP-60, March 1971, pp. 183-199.

[8] Grard, R., "Properties of the Satellite Photoelectron Sheath Derived from Photoemission Laboratory Measurements," Journal of Geophysical Research, Vol. 78, June 1973, pp. 2885-2906.

[9] Spangenberg, K., Vacuum Tubes, McGraw-Hill, New York, 1948.

[10] Bartlett, R.O., DeForest, S.E., and Goldstein, R., "Spacecraft Charging Control Demonstration at Geosynchronous Altitude," AIAA Paper 75-359, 1975, New Orleans, La.

[11] Mozer, F., "A Proposal to Measure Quasi-Static Electric Fields on the Mother/Daughter Satellites," Space Sciences Laboratory, University of California, Berkeley, Calif., Rept. 454, 1972.

[12] Gomer, R., Field Emission and Field Ionozation, Harvard University Press, Boston, 1961.

ACTIVE CONTROL OF SPACECRAFT POTENTIALS
AT GEOSYNCHRONOUS ORBIT

R. Goldstein[*]
Jet Propulsion Laboratory, Pasadena, Calif.

and

S. E. DeForest[+]
University of Alabama at Huntsville, Huntsville, Ala.

Abstract

The geosynchronous satellites ATS-5 and ATS-6 each carry plasma particle detectors (the University of California at San Diego (UCSD) auroral particle experiment) and two cesium (Cs) ion thrusters. The detectors have been used to determine spacecraft potential by observing shifts in the particle spectra and appearance of spurious particles at the spacecraft potential. During eclipse, negative potentials as great as 10 keV have been observed. Although the two spacecraft are markedly different in design and were separated by about 10° long for the tests reported, the potentials of the two, determined by the aforementioned method, agree surprisingly well while both are eclipsed simultaneously. Extensive tests of the effects of ion thruster beam, as well as thruster neutralizer electron source on spacecraft potential, have been performed. Similarly, the magnitude of the potential of the

---

Presented as Paper SA-39 at the American Geophysical Union Meeting, Washington, D.C., June 16-19, 1975. This work presents the results of one phase of research carried out at the Jet Propulsion Laboratory, California Institute of Technology, under Contract No. NAS7-100, sponsored by NASA. The authors are grateful to C. E. McIlwain, University of California at San Diego, for many helpful discussions in interpretation of the data. The ATS project office of Goddard Space Flight Center, in particular R. O. Bartlett, provided valuable support in carrying out the flight tests.

[*]Member of the Technical Staff
[+]Assistant Research Professor

spacecraft, charged negatively during eclipse, was reduced by operation of the ATS-5 thruster thermionic electron emitter. However, it was not always possible to clamp the potential to near zero by this technique.

Introduction

This paper describes and discusses some tests on the active control of the potentials of the geosynchronous satellites ATS-5 and ATS-6, and is a continuation of the study described previously by Bartlett, et al.[1] It should be pointed out that the ion thrusters used to control the spacecraft potential, as well as the charged particle detectors used to estimate the magnitude of the potential, were designed for other purposes; hence, a certain lack of flexibility in the experiments may be apparent.

Figure 1 (taken from Ref. 1) illustrates the basic configuration of ATS-5. It is a 184.2-cm-long, 146.3-cm-diam cylinder rotating at about 76 rpm. The two components of primary concern here are the filament neutralizer of the Cs-ion thruster, and the UCSD particle detectors. The filament is yttrium-doped tantalum, operating at about 1700°C. It is powered by a 3.5-V ac power supply and operates at spacecraft ground. Hence, the maximum energy (relative to the spacecraft) of electrons emitted is 3.5 V. At its normal operating

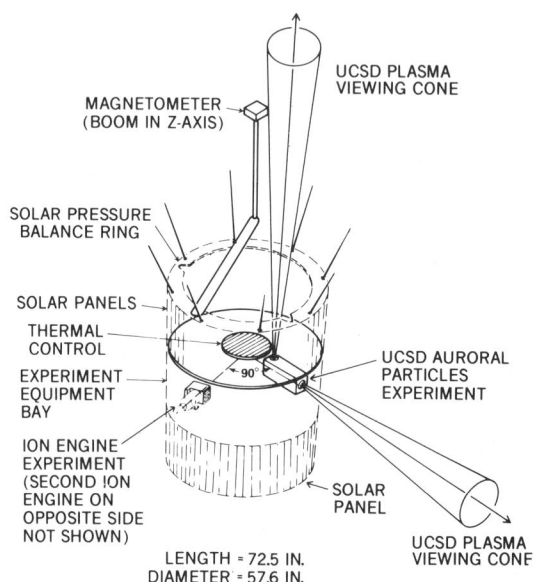

Fig. 1 ATS-5 spacecraft configuration.

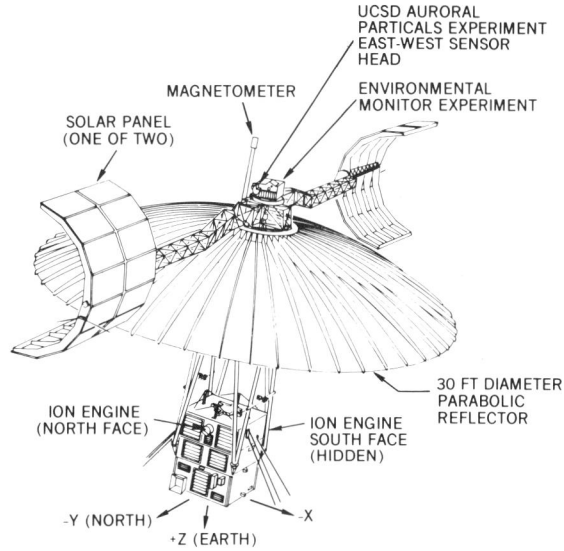

Fig. 2 ATS-6 spacecraft configuration.

temperature, the filament emission limits at a few milliamps. The thruster is located within a 5-cm-diam aperture in the spacecraft surface, with the filament recessed about 2.5 cm from that surface.

The UCSD experiment consists of two pairs of detectors, one pair viewing parallel to the spacecraft axis, the other pair perpendicular to it. Each pair consists of an electron and a proton detector covering an energy range, in steps, from 50 eV - 30 keV. Details of these detectors have been described by DeForest and McIlwain.[2]

The configuration of ATS-6 is illustrated in Fig. 2. The total end-to-end distance between the solar panels is about 15.5 m. The 9.14-m-diam mesh parabolic reflector separates the Earth viewing module (EVM) from the environmental measuring experiment (EME). The EVM is rectangular, about 160 cm high and 147.3 cm wide. Two electron-bombardment Cs-ion thrusters are mounted at the upper part of this module, one on the north-facing side and one on the south-facing side. The thruster axes point downward toward Earth at about 60° from horizontal. Each thruster is capable of producing up to about 115 mA in a well-collimated 550-eV ion beam. Neutralization is obtained with plasma bridge neutralizers. Additional details of thruster operation are given in Ref. 1 and the paper by Worlock, et al.[3]

The EME contains the UCSD particle detectors used to determine spacecraft potential. The detector system consists of electron and proton rate-counting pairs. One pair is in the north-south plane, and another is approximately in the east-west plane. The detectors have the capability of angular sweeping within their respective planes and cover the energy range of approximately 0.1 eV - 80 keV. Further details about their operation can be found in Refs. 1 and 4.

Spacecraft-Charging

The amount of charge-up of the spacecraft can be estimated by observing the shift in the particle flux-energy distribution measured by the UCSD detectors. For example, if the spacecraft is at some negative potential V, then the proton detectors will show no particles of energies less than V at that time. In addition, all particles of true energy less than V will have been shifted and appear as a spurious peak at V. Both daytime and eclipse charging events have been observed for both ATS-5 and ATS-6. During most periods in which both spacecraft are in eclipse simultaneously, it has been determined that they both charge to very similar voltages.[1] This is in spite of the fact that the spacecraft are of considerably different size, configuration, and material. In addition, ATS-5 is at about 105° W long, whereas ATS-6 was located at 94° W for these tests. (Both spacecraft are in near-equatorial orbits.) As a result of this general agreement in potential, it is possible to use one spacecraft as a "control", while actively modifying that of the other.

ATS-5 Discharge Tests

Because of the excessive satellite spin, the normal operation of the ATS-5 ion thrusters has not been possible. However, the neutralizer-filament electron emitter has been operated for 5-min intervals during the approximately 3-wk periods at the Fall 1974 and the Spring 1975 equinoxes, during which the spacecraft is eclipsed every night. In every one of the many cases thus far analyzed for which ATS-5 was charged negatively during eclipse, operation of the electron emitter resulted in a very rapid (<0.25-sec) drop in magnitude of the spacecraft potential. In most, but not all cases, this drop was followed subsequently by a slow increase in magnitude of the potential. In the other cases, the potential remained approximately constant while the filament was operating.

An example of the effect of operation of the neutralizer filament on the potential of ATS-5 is shown in Fig. 3 (taken from Ref. 1). This shows the history of spacecraft potentials

ACTIVE CONTROL OF SPACECRAFT POTENTIALS

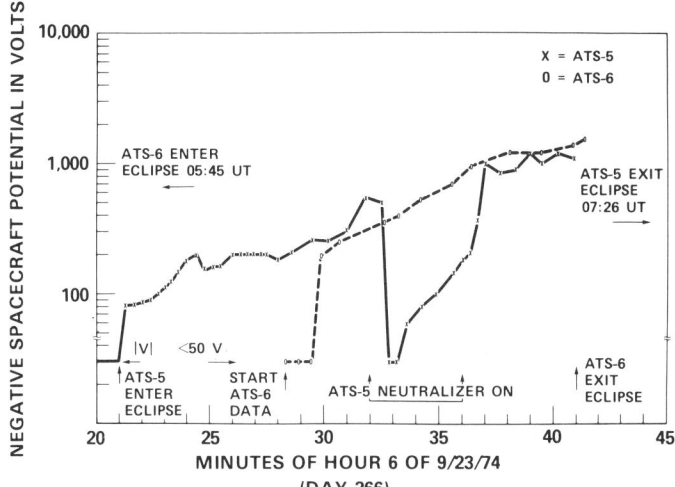

Fig. 3 ATS-5 and -6 potential history during eclipse.

for ATS-5 and ATS-6 during a portion of the eclipse period of Sept. 23, 1974. Before entering eclipse, the magnitude of the ATS-5 potential was $|V_5|$ <50 V. Upon entering eclipse at about 06:21 UT, V_5 suddenly jumped to nearly -100 V and then continued to become more negative. The data did not show ATS-6 entering eclipse, but, when they became available, showed the ATS-6 potential V_6 to follow very closely that of ATS-5. At 06:32, the ATS-5 neutralizer filament was turned on, and $|V_5|$ fell to <50 V. Shortly thereafter, $|V_5|$ began to increase, but at a much slower rate than $|V_6|$ did. At 06:36, the filament was turned off and V_5 jumped to a level in agreement with its previous rate of change and that of V_6. The two spacecraft left eclipse as shown in the figure.

Similar results for the Spring 1975 eclipse season are shown in a different manner in Fig. 4. Here an energy-time spectrogram for each satellite is shown for the period of eclipse on the night of March 20, 1975. The measured particle fluxes are given by a gray scale, higher fluxes being represented by the lighter areas. The upper pair is for ATS-6 (north-south plane), whereas the lower pair is for ATS-5 (perpendicular detectors). Energy is indicated along the ordinate axes, time in hours along the abscissa. Before entering eclipse, ATS-6 showed an appreciable flux of high-energy electrons and the peak of low-energy ($\lesssim$10 eV) electrons normally seen for this spacecraft. The latter are presumably photoelectrons and other secondaries. Very few low-energy protons were present. Upon entering eclipse, at about 05:50 UT, a marked change occurred in both the electron and proton count rates. The spacecraft rapidly charged negatively

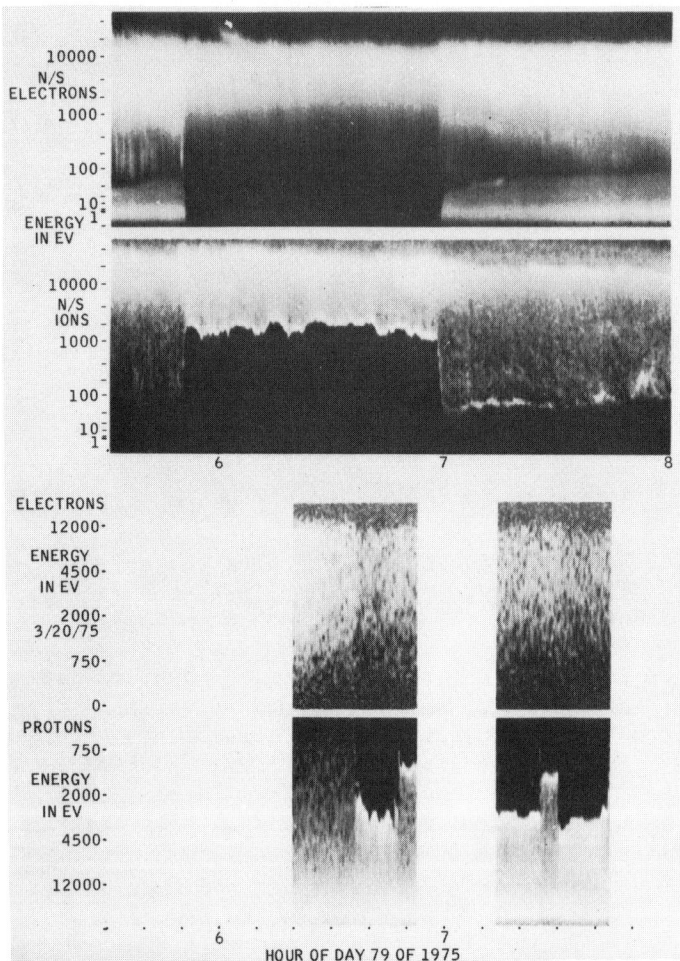

Fig. 4 Energy-time spectrograms for ATS-5 (lower pair) and ATS-6 (upper pair) during eclipse.

to a few thousand volts, as indicated by the shift of the proton spectrum. At the same time, the low-energy secondary electrons almost disappeared completely, and the higher-energy count rate also decreased. This is to be expected if a large fraction of the secondaries are photoelectrons. In addition, all electrons are repelled or retarded by the negative spacecraft potential.

At about 07:00 UT, ATS-6 exited eclipse and the secondary-electron flux reappeared. Note from the proton data, however, that apparently the high-energy electron flux had increased sufficiently to produce some daylight charging (~100 eV) at that time.

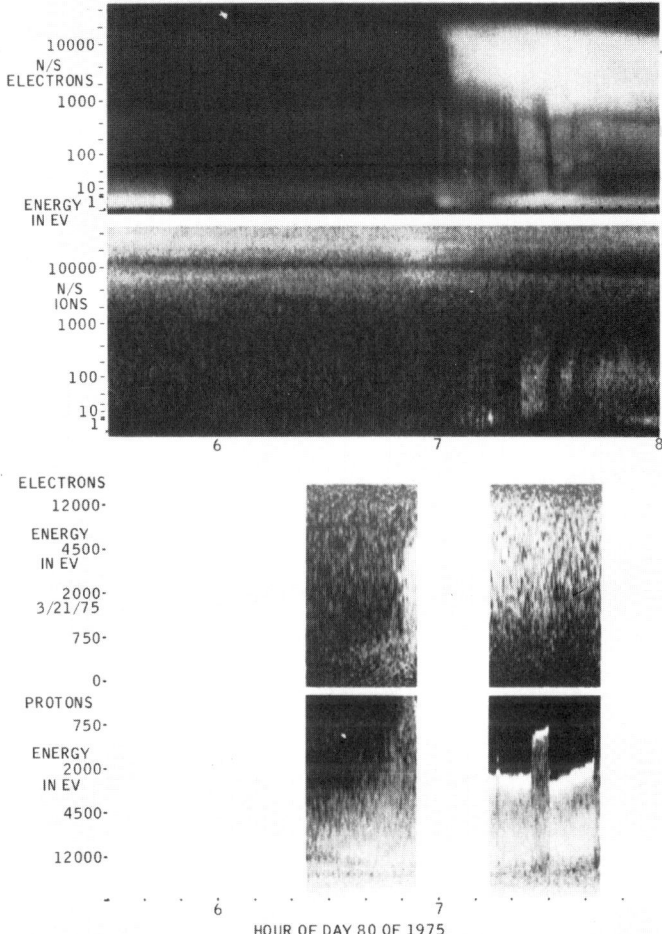

Fig. 5 Energy-time spectrograms for ATS-5 (lower pair) and ATS-6 (upper pair) during eclipse.

Meanwhile, ATS-5 entered eclipse at about 06:38 UT and rapidly charged negatively to a few thousand volts (in agreement with ATS-6). At about 06:47, the neutralizer filament was turned on, decreasing $|V_5|$ to several hundred volts. The potential then slowly became more negative. The white band is the result of missing data, which prevent observation of the filament turnoff. Another 5-min period of operation occurred at about 07:25 - 07:30.

Spectrograms for the next night, March 21, 1975, are shown in Fig. 5. In this case, very-low-particle fluxes were present before the spacecraft entered eclipse. The cutoff of low-energy secondaries was evident, however, when ATS-6 was in shadow. Since $|V_6|$ was less than a few volts at that point, the absence of secondaries was clearly due to only the absence of their major source, photo-emission, rather than their being repelled by a large negative potential. ATS-6 exited eclipse at about 07:00, and shortly thereafter an intense flux of high-energy electrons appeared. A slight amount (~10 V) of daylight charging may have resulted.

Similarly, ATS-5 entered eclipse and saw no measurable charging until (during the missing data) the high-energy electrons appeared; then $|V_5|$ ~ a few thousands volts. The 5-min operation of the electron source is seen at 07:25, considerably reducing $|V_5|$. Operation of the low-energy electron source on ATS-5 shows clearly the ability to reduce the amount of spacecraft-charging. However, details of its effects are not understood clearly at this time. In particular, although the filament is more than capable of supplying the current required to discharge ATS-5 completely, it has not been possible to maintain the spacecraft at near-zero potential during eclipse. There are two possibilities for this. First, if a negative potential well exists outside the satellite, as has been indicated for ATS-6,[5] the electrons may not be sufficiently energetic to leave the vicinity of the spacecraft. Secondly, the location of the filament within a recess may produce local space-charge effects which limit the emission current.

ATS-6 Discharge Tests

Satisfactory operation of both of the ATS-6 ion thrusters was achieved for limited periods in 1974. The south thruster was operated for approx 1 hr in July, and the north engine operated for 92 hr in October. Some results of the July test were reported previously.[1] Here we will discuss results of the October thruster operation.

Figure 6 shows a N-S particle spectrogram covering the period of day 292 (Oct. 19) during which the north thruster was turned on and reached stable operation. The periodic structure is a result of the angular sweeping of the detectors. (This sweeping ceased at about 09:45 UT.) The early part of the spectrogram clearly shows the low-energy secondary-electron peak. A significant flux of high-energy electrons is shown also. The absence of any apparent ion peak indicates a near-zero or slightly positive spacecraft potential. There is some indication of the possibility of charging in sunlight to several tens of volts negative between about 07:00 - 07:30 UT, however. Shortly thereafter (at 07:43) the thruster neutral-

ACTIVE CONTROL OF SPACECRAFT POTENTIALS

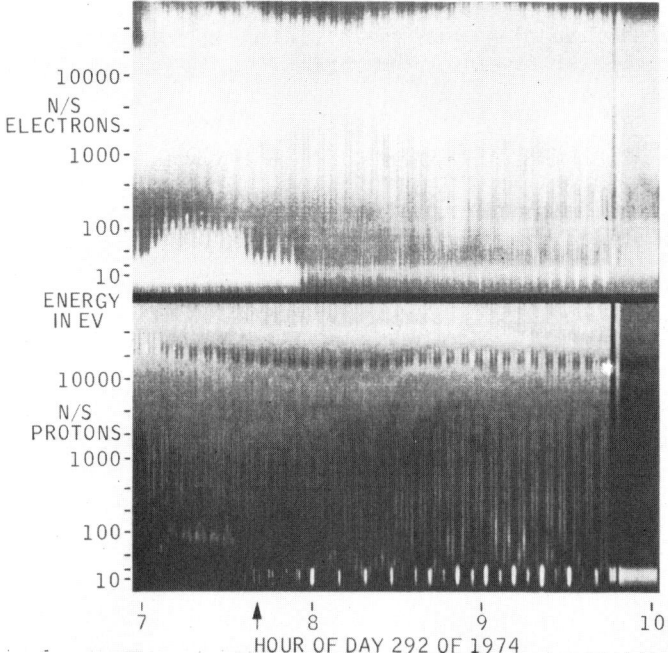

Fig. 6 ATS-6 spectrogram showing start of ion thruster operation.

izer discharge was ignited, creating a low-flow, low-energy neutral Cs plasma. Concurrently, the low-energy electron flux decreased, and a low flux of ions appeared at about 10 eV. Finally, at 08:01 the main discharge struck, and a beam was formed. The sudden drastic decrease in the secondary electrons is evident, as is the increase in low-energy ions. By 08:52 the thruster was stabilized at 110-mA beam current and remained very nearly constant until termination of the test on day 296.

Appearance of the low-energy ion peak can be attributed to two sources. First, operation of the thruster would tend to clamp the spacecraft to a low negative (~5-10 V) potential characteristic of the coupling of the thruster to the surrounding plasma. This would result in the proton peak occurring at that energy, as is typically the case when the spacecraft is charged negatively. Secondly, thermal energy Cs-ions created by charge exchange in the beam[6,7] would be attracted by that small negative potential on the spacecraft and would appear at the energy corresponding to that potential. This is indicated also by the dependence of the low-energy ion-count rate on thruster beam current, which will be discussed below in more detail.

The sharp decrease in the secondary-electron count rate must be attributed to an increased rate of loss of these particles from the spacecraft vicinity, since presumably their production rate remained constant. This must be an indication that the potential barrier keeping them at the spacecraft[5] had been "breached", or drastically modified, by the thruster ion beam. Note that the large physical separation between the particle detectors and the thruster (~5 m) indicates that the effect on the barrier must have been significant. This point will be taken up again below.

Throughout the entire 90-hr period of steady-state operation of the thruster, the spacecraft was maintained at about −10 V. This includes occasions on which the high-energy electron flux most likely would have been sufficient to charge the spacecraft several hundred volts negative in the absence of the ion beam.

The dependence of the low-energy electron and ion count rates on thruster beam current is shown in Fig. 7. (The history of thruster beam current in the lower portion of Fig. 7 was kindly supplied by E. L. James and R. M. Worlock of Xerox

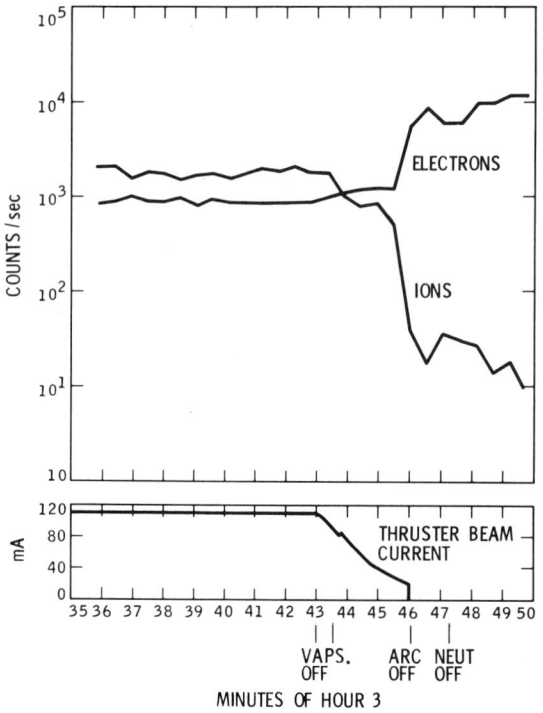

Fig. 7 ATS-6 low energy particle count rates and thruster beam current history during thruster turn-off.

Electro-Optical Systems.) In the upper portion, the count rates of the low-energy peaks (~10 eV) are plotted as a function of time at the end of the test when the thruster was turned off. The turnoff period was chosen to avoid the complications of interpretation of the periodic structure due to the angular sweeping of the detectors, which were in effect during the thruster turnon. The lower portion of Fig. 7 shows the history of thruster beam current during that same period. As mentioned previously, the beam current was steady at 110 mA for the entire length of the test. The electron and ion count rates were constant over most of the test, as shown in the early part of the figure, but occasionally increased when ambient conditions changed drastically. (Note, however, that there was no apparent change in the energy of these low energy peaks.) At 03:43:00 and 03:43:30, the anode and cathode vaporizers, respectively, were turned off. The resulting decrease in Cs flow to the thruster is reflected in the decreasing beam current. There was also an immediate decrease in ion counts and an increase in electron counts. These changes continued at a relatively low rate as the beam current fell, until, at about 03:45:30, the rates of change increased suddenly. This corresponds to about 30-mA beam current. Finally, at 03:46 the Cs flow was too low to maintain the arc discharge within the thruster, and it extinguished, cutting off the beam completely. For slightly more than 1 min beyond that time, only the (decaying) neutralizer discharge was present. A very small effect on the count rates is observable during this last period.

Discussion

The dependence of the ion count rate on beam current indicates clearly that a large fraction of these particles must originate in the beam, i.e., are probably low-energy Cs-ions produced by charge exchange and attracted back to the spacecraft.[6,7] The ion cyclotron radius for thermal Cs-ions is of the order of several kilometers in the ~1.5-mG field at the spacecraft. However, since the ion beam was roughly along the magnetic field direction, some ions could have reached the detectors over a path perhaps as short as a few tens of meters. In addition, differential charging between the various nonconducting surfaces of the spacecraft complicate the electric field distribution and therefore the ion trajectories between the thruster and detectors.

Thus, these thermal ions must be spread over a very large region. This is likely to be the reason for the strong effect of the thruster on the secondary-electron count rate, previously described. The total charge-exchange-ion flux should be proportional to the square of the beam current (for constant propellant utilization efficiency). Thus, any direct effect of the charge exchange ions on the spacecraft environment should have a strong dependence on beam current also. But this is exactly what appears to be the case in Fig. 7, where the measured electron and ion count rates break sharply at ~30-mA beam current.

This is, of course, only a qualitative argument; but the evidence is strong that it applies. For example, it is not possible to separate out the extent of direct modification of the barrier by the primary thruster beam itself. However, most of this beam is contained within a 20° half-angle, and hence its effects are probably more local in nature than the charge exchange cloud. The ions attracted directly to negatively charged insulator surfaces will tend to neutralize those surfaces. This process may also have an important effect on the potential barrier.

The relatively small effect of the neutralizer on the measured ion count rate shown in Fig. 7 appears to be inconsistent with the relatively larger change measured during the initial turnon on day 292. The reason for this is not clear, but may involve differences in operating conditions of the neutralizer not readily apparent from the telemetry.
The relative change in secondary electron count rate is similar, however, in the two cases. That this change is relatively small indicates that the neutralizer has little effect on the potential barrier, although some potential "clamping" effect may occur.

Summary and Conclusions

The ATS-5 tests show that a simple electron emitter is capable of at least partially reducing the magnitude of the potential of the spacecraft that has been charged negatively by the environment. The exact nature of the interaction between the electrons released from such a device and the local environment is not well understood, however. The indication is that the electrons must be ejected at sufficient energy to overcome the potential barrier surrounding the spacecraft in order to discharge it completely. In addition, it does not appear that such electrons in any way would alleviate the potential barrier. Hence, low-energy "natural" electrons outside the barrier still would be prevented from reaching a detector on the spacecraft. Resolution of this matter must await a more complete explanation of the barrier phenomenon than is presently available.

Of course, insulator surfaces cannot be discharged with a simple electron emitter. In fact, by discharging the spacecraft frame while allowing insulators to remain charged may result in excessive electrical stress on those insulators.

On the other hand, in addition to clamping the spacecraft to a small (~10 V) negative potential for a wide range of ambient plasma conditions, the ATS-6 ion thruster did show a marked effect on the potential barrier as well. In this case, the flux of both high-energy primary ions, as well as low-energy charge exchange ions, must exercise a space-charge neutralization effect which the electron gun alone cannot achieve. Hence, any spacecraft that is subject to the potential barrier phenomenon would require some type of neutral plasma or ion plus electron guns to allow measurement of low-energy particles.

References

[1] Bartlett, R. O., DeForest, S. E., and Goldstein, R., "Spacecraft Charging Control Demonstration at Geosynchronous Altitude," AIAA Paper 75-359, March 1975, New Orleans, La.

[2] DeForest, S. E. and McIlwain, C. E., "Plasma Clouds in the Magnetosphere," Journal of Geophysical Research, Vol. 76, June 1971, pp. 3587-3611.

[3] Worlock, R. M., James, E. L., Hunter, R. E., and Bartlett, R. O., "The Cesium Bombardment Engine North-South Stationkeeping Experiment on ATS-6," AIAA Paper 75-363, March 1975, New Orleans, La.

[4] McIlwain, C. E., "Auroral Electron Beams Near the Magnetic Equator," Physics of the Hot Plasma in the Magnetosphere, edited by B. Hultqvist and L. Stenflo, Plenum Publishing Co., New York, 1975.

[5] Whipple, E. C., Jr., "Observation of Photoelectrons and Secondary Electrons Reflected from a Potential Barrier in the Vicinity of ATS-6," Journal of Geophysical Research, Vol. 81, February 1976, pp. 715-719.

[6] Kemp, R. F., Beynon, J. C., Luedke, E. E., and Hall, D. F., "Effects of Cesium Ions and Cesium Vapor on Selected ATS-F Samples," AIAA Paper 73-1099, Oct. 31-Nov. 2, 1973, Lake Tahoe, Nev.

[7] Worlock, R. M., Trump, G., Sellen Jr., J. M., and Kemp, R. F., "Measurement of Ion Thruster Exhaust Characteristics and Interaction with Simulated ATS-F Spacecraft," AIAA Paper 73-1101, Oct. 31-Nov. 2, 1973, Lake Tahoe, Nev.

Chapter 3—Spacecraft Materials; Response to the Arcing Process

A definitive laboratory test program is necessary in order to a) study the response of materials, components, and assemblies to charge buildup and arc breakdown; b) experimentally establish the interaction and the mechanism for the substorm-induced anomaly; and c) provide a quantitative basis for designing future spacecraft so as to eliminate their susceptibility to substorm-related phenomena.

The susceptibility of a spacecraft to high-voltage differential charging depends on the selection of spacecraft materials and on the dynamics of the arcing phenomena under standard environmental conditions. It is not possible to calculate voltage stress levels or to design a spacecraft so as to minimize these levels without knowledge of the bulk and surface resistivities of materials, the photoconduction magnitudes, the secondary emission ratios, and the photoemissive currents for both dielectrics and conductors. Unfortunately, some of these parameters have not been determined for the material commonly used in spacecraft design. For example, in the first paper, Hoffmaster and Sellen have measured the bulk resistivity of Kapton and find that the resistivity varies with the applied electric field stress level and light irradiation. The characteristics of the discharge process and a discussion of the discharge mechanism are presented in that paper and in the series of four papers that follows.

The last two papers address the problem of spacecraft design and test so as to reduce its susceptibility to environmental effects. The basic approach is to eliminate all capacitive configurations that can accumulate charge from the environment by either grounding or strapping the proper elements and to ground and shield circuits so as to reduce their susceptibility to arc generated EMI. A spacecraft test verification program also may be necessary to assure that all capacitive configurations have been eliminated or reduced and that the spacecraft circuitry is not susceptible to arc generated RFI.

SPACECRAFT MATERIAL RESPONSE TO GEOSYNCHRONOUS SUBSTORM CONDITIONS

D. K. Hoffmaster[*] and J. M. Sellen Jr.[+]

TRW Defense & Space Systems, Redondo Beach, Calif.

Abstract

A series of measurements of metal-to-metal arc discharges and bulk resistivity of Kapton at high stress levels has been made. Arc currents in excess of 1 ka and with rates of rise in excess of 20 Ga/sec have been observed, with periods of less than 10^{-7} sec from breakdown to full current, and $\sim 3 \times 10^{-7}$ sec for the total current pulse. Processes in arc formation, energy transfer, and arc motion have been discussed, with limiting factors noted, and additional directions for analysis and testing noted. The bulk resistivity of Kapton, under levels of electric stress from 0.1 to 1.6 MV/cm, has been examined in both uv-absent and uv-present environments. For the most severe stress, resistivities of the order of 10^{14} Ω-cm are observed, some four orders of magnitude lower than the nominally quoted figure of 10^{18} Ω-cm bulk resistivity for Kapton. When uv light at 2536 Å is directed on the sample, and for high stress, resistivity loss by a factor of roughly 4 is observed. The uv light during these exposures

---

The interest of A. Haeff, G. Inouye, A. Krausz, and A. Rosen during the course of the metal-to-metal arc experiments and analyses is gratefully acknowledged. G. K. Komatsu and N. L. Roy assisted in the calibration of probes and in the reduction of data. For the bulk resistivity measurements and analysis, the interest of G. T. Inouye, A. Rosen, and G. K. Komatsu of TRW Defense & Space Systems, W. Capps and D. P. Cauffman of Aerospace Corporation, and S. E. DeForest of the University of California, San Diego, in the conduct of the experiments and in the discussion of the program results is gratefully acknowledged.
[*]Senior Laboratory Technician.
[+]Senior Scientist.

is significantly below solar uv intensity, so that even larger increases in conductivity would be expected for samples exposed to solar uv.

1. Introduction

Spacecraft in geosynchronous orbit and under magnetic substorm conditions acquire significant levels of charge deposition.[1-5] Since spacecraft surfaces are composed of a variety of materials, both insulating and conducting, and since the response of these materials to the substorm environment will differ from one area to another as material properties vary, the extent of charge-up also will vary, leading to "differential" charging and accompanying high values of electric stress.[6,7] These stress conditions lead to current leakage through the material, which is dependent on the bulk resistivity of the material, and, beyond certain threshold values, to electrical breakdown or arcing.

Two forms of electrical breakdown have been identified in charge relocation processes for spacecraft materials. These are metal-to-metal arcs following breakdown of an intervening dielectric material, and dielectric-to-metal arcs, also following dielectric breakdown. This paper will describe measurements of the bulk resistivity of Kapton at electric stress levels from 0.2 to 1.6 Mv/cm and measurements of metal-to-metal arcs in which current flow occurs following a surface breakdown of dielectric material. Metal-to-metal arcs, involving bulk breakdown of the intervening dielectric and dielectric-to-metal arcs also have been observed and are discussed in this paper and in other papers in this volume.[8,9]

2. Electron Swarm Tunnel Facility

The measurements described in this paper were performed in the 2- x 4-ft electron swarm tunnel (EST) facility illustrated in Figs. 1 and 2. Electrons generated by a filament are accelerated in a linear Pierce gun that is defocused to provide beam divergence and larger dielectric sample coverage. Acceleration voltages range to 20 kv. Beam current densities are nominally in the range from 10-1000 na/cm^2, simulating from 1-100 times substorm electron deposition current densities: beam current density profiles are determined by the movable Faraday cups. Chamber pressure is maintained at $\sim 10^{-6}$ torr.

The magnetic field coils, when activated, cause the beam to deflect away from the target and onto the chamber walls.

SPACECRAFT MATERIAL RESPONSE 187

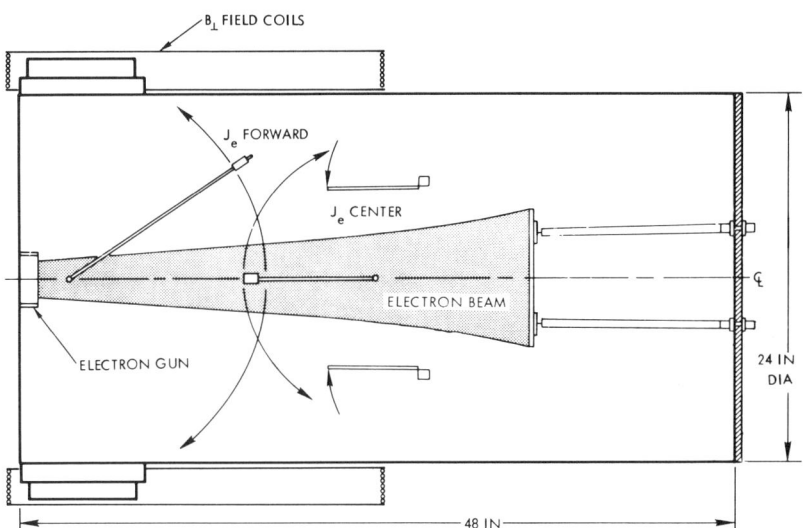

Fig. 1 Top view of 2- x 4-ft electron swarm tunnel illustrating sample placement and protecting movable doors (shown opened to allow sample exposure). Sample charge-up causes illustrated beam divergence.

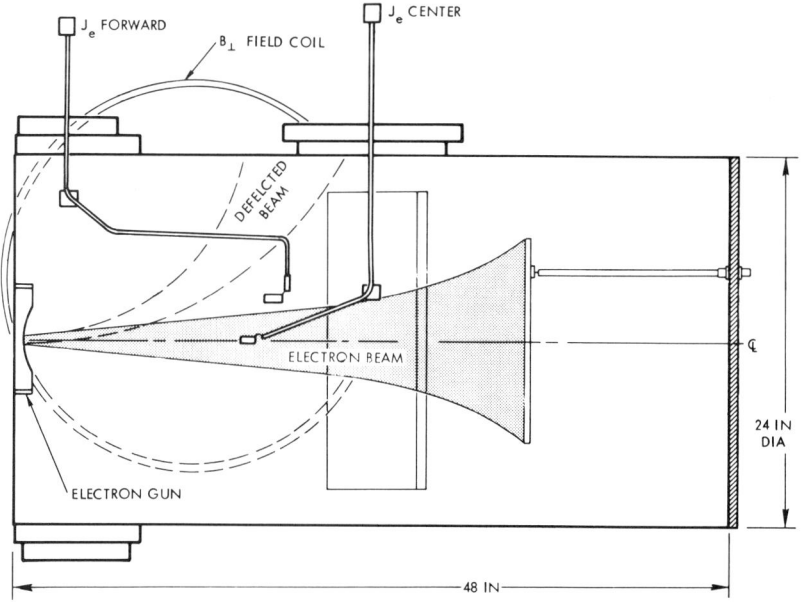

Fig. 2 Side view of 2- x 4-ft electron swarm tunnel illustrating beam, sample, doors, and probes.

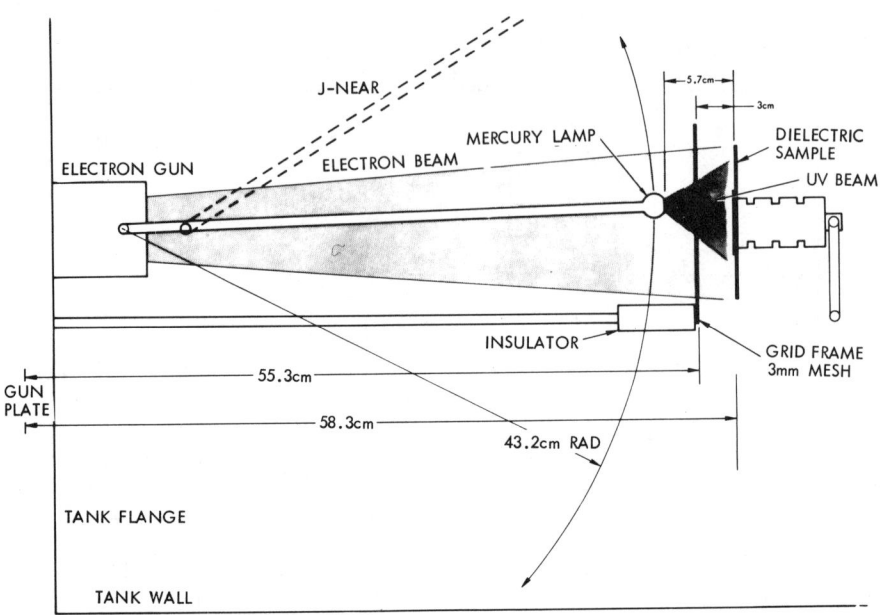

Fig. 3. Electron swarm tunnel movable uv light source and surface potential stabilization grid.

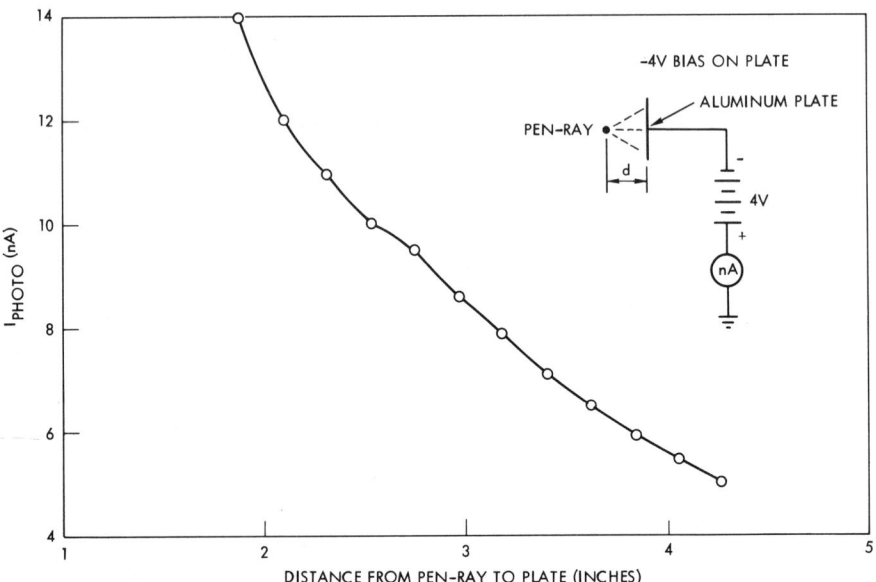

Fig. 4. Photoemission current from aluminum as a function of uv source-to-plate separation distance.

By reducing the magnetic deflection and allowing the electrons to deposit on the surface of a previously uncharged dielectric, an integral of the total dielectric surface charge may be obtained, and surface potential and electric stress in the bulk material may be calculated.

The placement of the uv source within the EST is illustrated in Fig. 3. The radiation from the mercury arc lamp shown there is primarily at 2536 Å, with minor line strength at 1849 Å. By movement of the lamp support arm and the sample holder, the source-to-sample distance may be varied, and hence, the intensity of uv deposited on the sample. Figure 4 illustrates photoelectron current from an aluminum surface as a function of the separation distance between the EST uv source and the plate. Using the total aluminum surface area, 100 cm^2, photoelectron current density is $\sim$0.14 na/cm^2 at 5 cm separation. Since solar uv causes photoemission from aluminum at $\sim$5 na/cm^2, the present EST uv source is not at solar uv equivalency, and in-orbit diminutions of dielectric bulk resistivity from uv deposition would be expected to be much larger than those observed in the laboratory photoconduction experiments.

3. Metal-to-Metal Arc Configuration for EST Measurements

For the metal-to-metal arc measurements the sample is located behind movable metal doors that protect the materials from charge-up during periods of e-beam calibration. Following e-beam calibration, the electrons are deflected onto the chamber floor and the protective metal doors are opened. To initiate charge-up and the subsequent metal-to-metal arc, the magnetic deflection is pulsed "off," allowing the EST beam to impact on the target sample. The electron energy in the swarm tunnel was set at 10 keV with current densities of the order of 100 na/cm^2. The cross sectional area of the e-beam, at the sample location, was approximately 10^3 cm^2 which was also the exposed sample area.

Figure 5 shows a cross section of the VDA/Kapton/metal sample placement. In these measurements, the forward-facing (and electrically isolated) metal layer is exposed directly to the electron stream, whose deposition causes the metal to move to a negative charge-up potential with respect to the second (grounded) conducting layer at the rear. It may be noted that metallized dielectric thermal materials are most commonly mounted on a spacecraft surface with the insulation facing outward. However, conduction does occur in dielectrics and, thus, an interior (and electrically isolated) metal layer may acquire charge and potential relative to another (grounded) metal surface in the same manner as occurred in the EST. There are, also, capacitive voltage dividing processes that cause isolated internal metal surfaces to differentially charge with respect to other metal surfaces. The configuration in Fig. 5 thus is a relevant metal and dielectric array.

The cross section in Fig. 5 illustrates one of three "return loop" configurations used in the arc measurements. Figure 6 illustrates all three of the configurations and also provides a front view of the metal/dielectric sample. The discharge from the isolated and electrically charged forward facing metal layer occurs following surface breakdown across a dielectric gap of approximately 5 mm. This discharge current then proceeds, for samples A and B, along a metallized strip around the edge of the dielectric sheet, through a single-turn "primary" loop, and to the rear face metal layer (at ground potential). The primary loop dimensions for samples A and B

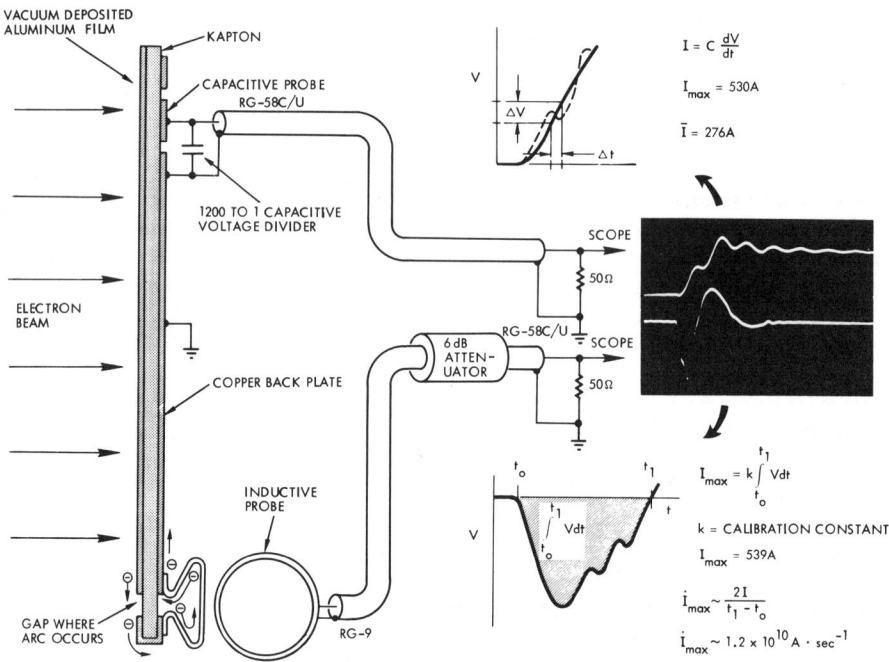

Fig. 5. Cross section of VDA/Kapton/metal sample illustrating probe placement, probe output traces for a given metal-to-metal arc, and derived peak arc current and arc current time derivative.

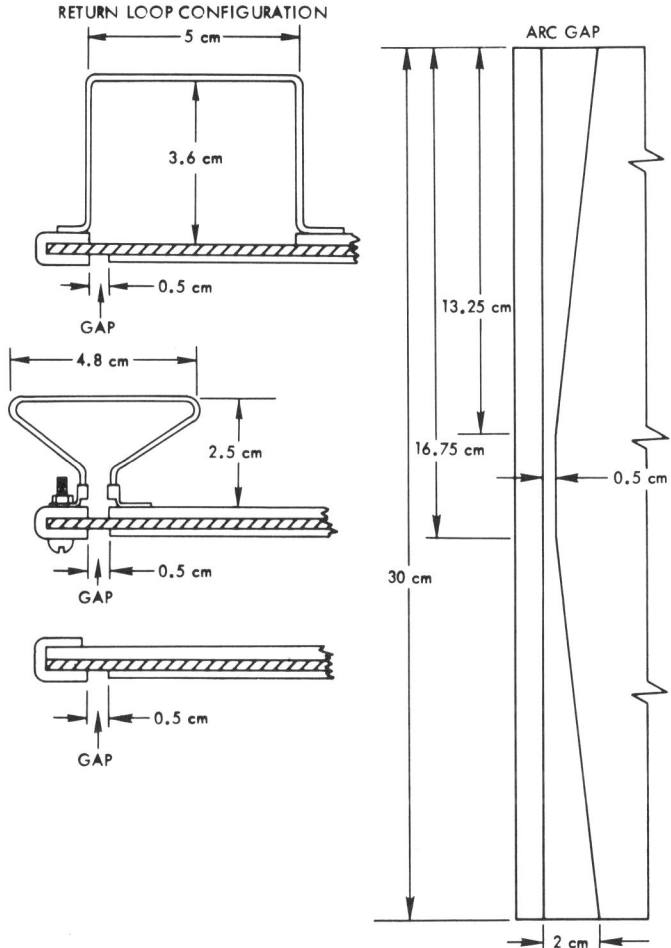

Fig. 6. Current return configurations, dielectric/metal cross sections, and dielectric gap for samples A, B, C, and front view of metal surface/dielectric gap shaping.

are given in Fig. 6 with the reduction in area of sample B, compared to A, to reduce inductance for the single turn, and hence, to permit the maximum arc current and arc current time derivative. Current in the primary loop was sensed by a single-turn secondary of 4 cm diameter.

For sample A, arc diagnosis was limited to the illustrated single turn transformer. Sample B, however, possessed capacitive probes to the forward metal layer and, thus, provided a measurement of time-rate of charge of the forward-face potential. By appropriate integrals and derivatives of these inductive and capacitive probes, in situ current, and voltage calibration pulse data, and through the known metal-to-metal capacitance, two measured values of arc current and current-time derivative were obtained. The consistency between these two approaches indicated that capacitive probe measurements were sufficiently accurate; for sample C, the inductive probe "primary loop" was eliminated, thus providing minimum possible inductance to the arc discharge and maximizing arc current, I_A, and dI_A/dt.

4. Measured Properties of Metal-to-Metal Arcs

The current pulse from the forward to the rear metal face following dielectric surface breakdown will depend upon several factors, among which are the front-to-back charge-up voltage prior to discharge, the lateral point along the dielectric gap at which breakdown initiates, and cumulative alterations of both conducting and dielectric surfaces following discharges. In addition, resistance, capacitance, and inductance of the metal/dielectric/metal will determine the time rate of transfer from initially stored energy to the arc, whereas arc formation and motion will influence the arc impedance. A discussion of these variations is presented in Sec. 7.

In the EST tests, the e-beam moved from an initially deflected position to impact on an initially uncharged, electrically isolated front face. Charge buildup, and potential buildup on the front face followed and was determined in a series of tests by measurements of front-to-back displacement currents and the known front-to-back capacitance. These tests revealed that front surface potentials increased by several thousand volts before dielectric surface breakdown occurred. The voltage buildup varied depending upon the number of previous discharges, and although some stabilization and triggering processes might have been devised to narrow the range of breakdown voltage, no such process is present for metals and dielectrics exposed to the space environment; for similitude, EST breakdowns were allowed to trigger "naturally."

Figure 7 illustrates peak values of discharge current, I_A, and peak values of dI_A/dt for a series of discharges on sample A. Indicated on that figure is the number of the breakdown, and it appears that initial breakdowns yielded larger currents and time rates of change of current than subsequent arcs. Several factors may be responsible for this behavior. There may be removal of dielectric surface material, or in

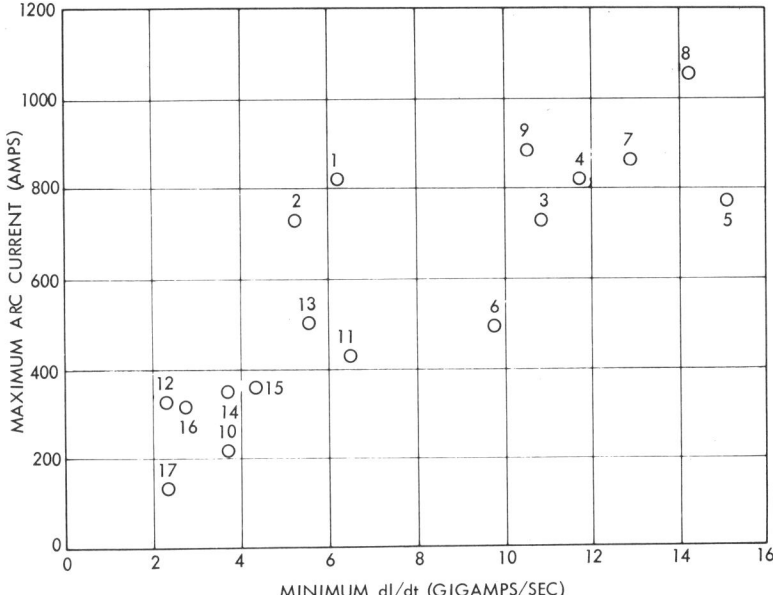

Fig. 7. Maximum metal-to-metal arc currents and minimum value of $(dI_A/dt)_{max}$ for 17 arcs observed on sample A (number on data point indicates arc sequence).

situ alteration of that material, or deposition of other (and perhaps, conducting) materials. (Significant quantities of vacuum deposited aluminum (VDA) are observed to be vaporized in the discharge process.) Another possible process of importance could be that the removal of VDA results in a now lengthened gap, and that the breakdown across this gap produces a higher impedance arc than observed initially, with correspondingly lowered current.

For sample A, peak arc currents of $\sim$1070 A and peak current growths of $\sim$1.5 x 10^{10} A/sec were observed. For such arcs and for a calculated primary return loop inductance of $\sim 10^{-7}$ H, it would appear that voltage drops of $\sim$2000 V could result from this $L(dI_A/dt)$ term. Since return loops on actual spacecraft materials under breakdown in substorms would not have this "primary" turn area or inductance, it was decided to reduce, as much as possible, the primary loop area; sample B was so constructed. In addition, capacitive probes were added as an initial step to the ultimate removal of the inductive loop.

Sample B, on exposure, demonstrated arcs of essentially the same magnitude as sample A, indicating that changes in return loop area had not resulted in significantly larger discharge currents. Figure 5 has illustrated the signals from inductive and capacitive probes. For the largest observed arc (prior to ultimate sample deterioration), peak I_A = 1030 A and peak dI_A/dt = 1.4 x 10^{10} A/sec. Peak current was obtained ~100 nsec after initiation of breakdown, and peak dI/dt occurred ~50 nsec after the initiation. Total time for the discharge was ~300 nsec. Although these results were not significantly different from those of sample A, the agreement of both inductive and capacitive probes allowed the removal of the inductive loop in the subsequent sample C.

Sample C, upon its initial charge-up and discharge exhibited a peak current of 1440 A, a peak current rate of rise of 2.5 x 10^{10} A/sec, a total front-to-back charge transfer and potential shift of 200 μcoul and 4700 V, and a total VDA vaporized area of ~1 cm^2. Following this initial charge-up and breakdown, only minor arcs were observed, leading to speculation that vaporized VDA may have been deposited over the intervening dielectric surface with a resulting diminution of breakdown potential and current. Irrespective of process, however, these initial arc currents and current time derivatives are large enough to raise concerns for the effects that such an arc could impose if it had occurred on a spacecraft. It is not apparent, moreover, that this presently observed arc represents a maximum possible discharge condition, and discussion in Sec. 7 will consider limiting processes in arc growth, noting where still larger I_A and dI_A/dt than those thus far observed might be generated.

5. Determination of Surface Potential, Electric Stress Level, and Bulk Resistivity at Fixed Electron Swarm Energy

5.1 Dielectric Sample Configuration and Placement

For bulk resistivity measurements, dielectric samples were located at the midplane of the 2- x 4-ft EST, and beam-current density profiles were taken prior to each exposure to assure adequate electron deposition over the sample surface. Figure 8 illustrates the Kapton dielectric sample. The 0.002 in (0.005 cm) Kapton foil was placed so that electrons from EST beam would deposit on the front (insulating) surface. Drainage currents through the central region of 7.1 x 7.1 cm were collected by a vacuum deposited aluminum (VDA) film on the rear surface. An outer VDA guard ring (10.2 x 10.2 cm outside, 9.1 x 9.1 cm inside) prevented surface leakage currents from the front surface to the central rear collector. During ex-

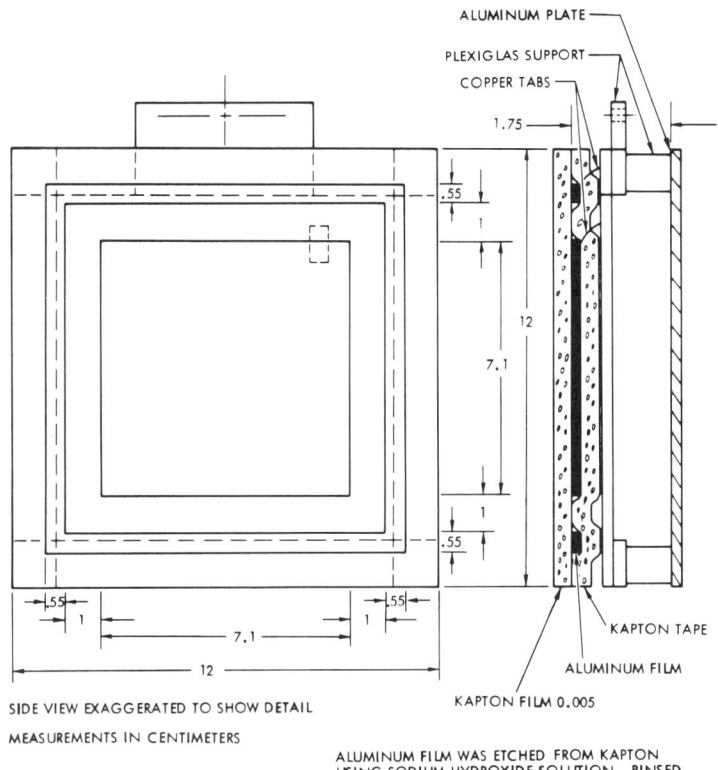

Fig. 8. Kapton dielectric sample with aluminum film collector and guard ring.

periments, guard ring current was measured with a low impedance circuit so that guard ring potential remained at $V \sim 0$. Currents to the central collector were measured with an HP 425A microammeter, followed by a Mosely 2D2 X-Y plotter.

5.2 Dielectric Surface Potential

The deposition of an energetic electron into a dielectric material results in a negative charge at some penetration depth below the surface and possible release of secondary electrons primarily from regions of the material near the surface. If secondary emission is initially less than unity, a net buildup of negative charge results, and the surface acquires a potential which is negative with respect to that surface potential which existed prior to energetic electron arrival. Electrons arriving at the (now) negatively charged surface will be decelerated by the space charge fields of the accumulated charge and will enter the material with an energy less than the original acceleration energy. An ultimate balance condition (for drainage currents that are small compared to bombarding currents) is that surface potential at $-V_s$ results in electron impact energy $e(V_o-V_s)$, where V_o is electron gun acceleration voltage, such that effective secondary emission, δ_{eff}, is unity. For this impact energy, the deposition of one electron results in the release of one electron, charge accumulation rate drops to zero, and the surface attains a steady state potential. This condition is described as a surface equilibration at the "second crossover potential" of the secondary emission curve (see Ref. 6, Fig. 6).

Because of the complexity of secondary emission and charge deposition in a dielectric, a convenient procedure for determining surface equilibration potential is to allow the electron beam to deposit on an initially uncharged dielectric and to integrate displacement current at the rear surface. This determines surface charge density, σ_e; from known dielectric capacitance per area, dC/dA, front-to-back potential difference is determined. With the rear (collecting) surface at $V = 0$, front surface potential, in volts, is given by

$$V_s = \sigma_e/(dC/dA) \qquad (1)$$

where σ_e, in C/cm^2, will be negative for electron accumulation, and dC/dA is in F/m^2. For Kapton, the dielectric constant is 3.5, and for a film thickness of 5×10^{-3} cm, $dC/dA \sim 0.6$ $\mu F/m^2$ (60 pF/cm^2).

Figure 9 illustrates the current from the rear collector as a function of time following the deposition of a 6.0 keV electron beam on the front surface. In the period from $t \sim 10$ sec to $t \sim 100$ sec, the irregular behavior of the current is a result of the step-wise movement of the beam from its deflected state to a nondeflected (target directed) state. From $t \sim 100$ sec to $t \sim 250$ sec the diminishing current is the result of diminishing rate of charge-up on the front surface. Over the period from $t \sim 10$ to $t \sim 250$ sec, the observed signal is principally a displacement current, but, for continued exposure, the front surface reaches a steady state potential, displacement current diminishes to zero, and the remaining steady state signal is a conduction current through the bulk of the dielectric.

In performing an integral of the displacement current to determine V_s, the exact cutoff point is not evident, since the

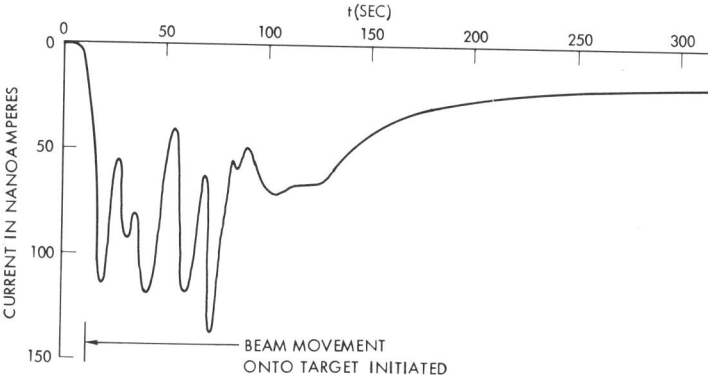

Fig. 9. Displacement and conduction current to aluminum film collector following deposition of 6-keV electron stream onto Kapton surface.

ultimately observed signal consists of both displacement and conduction currents. The calibrations of integrated collector charge and front-surface potential as a function of time were made assuming the entirety of the signal as displacement current and using total sample capacitance (50 cm^2) as 3000 pF. If the cutoff were performed at 200 sec, a front surface potential of -3820 V results, whereas a cutoff at 250 sec yields V_S = -4200 V. Within a possible error of 200 V, it would appear that the front surface acquires a negative potential of -4000 V. Since electron acceleration voltage is 6000 V, electron energy at impact in steady state equilibration is 2000 eV, and effective secondary electron emission is unity at that impact energy.

An alternate method of front surface equilibration potential utilizes the movable grid (Fig. 3). If the grid is moved into the electron stream, and near the source, and if V_g = 0, then beam electrons continue to move to the dielectric and deposit, secondary electrons are emitted and move away from the surface, and the front surface potential remains at its earlier steady state charge-up value. If grid potential is made increasingly negative, however, there is a point at which the grid becomes negative with respect to the surface; secondary electron emission is suppressed and the front surface potential moves in the negative direction. Evidence of a negative shift in surface potential is a displacement current surge (following the grid potential adjustment) and, in the ensuing new steady state conditions, increased levels of conduction current. The determination of surface equilibration potential in an electron

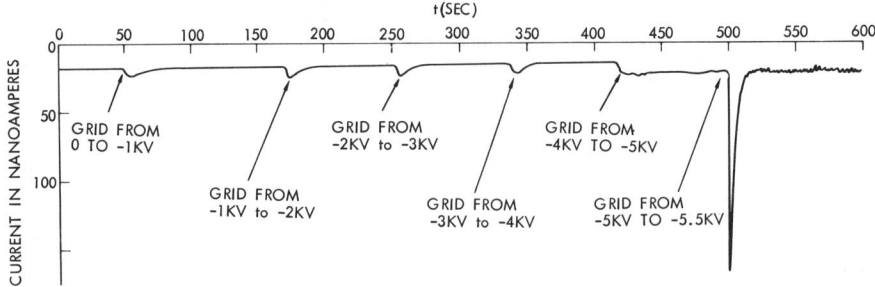

Fig. 10. Displacement and conduction current to aluminum film collector as a function of surface stabilization grid potential for steady-state 6-keV electron stream onto Kapton surface.

stream, then, is at the first appearance of a displacement current surge with subsequent elevated conduction level.

Figure 10 illustrates the rear collector current signal with the grid in place ∿3 cm from the dielectric surface, and as the grid potential is varied. For movement of the grid from 0 to -1000, -1000 to -2000, -2000 to -3000, and -3000 to -4000 V, only modest displacement current surges are observed, and there are no significant shifts in conduction current. As the grid is moved from -4000 to -5000 V, however, a very perceptible rise in conduction signal is observed; for grid potential moving from -5000 to -5500 V, a comparatively large displacement current surge occurs.

The drainage current and displacement current signals in Fig. 10 tend to confirm the earlier conclusion of a surface equilibration potential from electron beam deposition alone of ∿-4000 V. Because of the comparatively open structure of the grid employed, however, the method is not precise, and principal emphasis will be given to the result derived by the integral of the displacement current.

5.3 Electric Stress Levels

For the front-to-back potential difference of -4000 V, the electric field is calculated to be 8×10^5 V/cm for a film thickness of 5×10^{-3} cm. In the present (and comparatively qualitative) approach to dielectric behavior, the 0.8 MV/cm figure will be used as the applied E to produce the observed drainage.

Possible effects of space charge are illustrated in Fig. 11. For an extreme case of electron density of $\sim 2 \times 10^{14}/cm^3$,

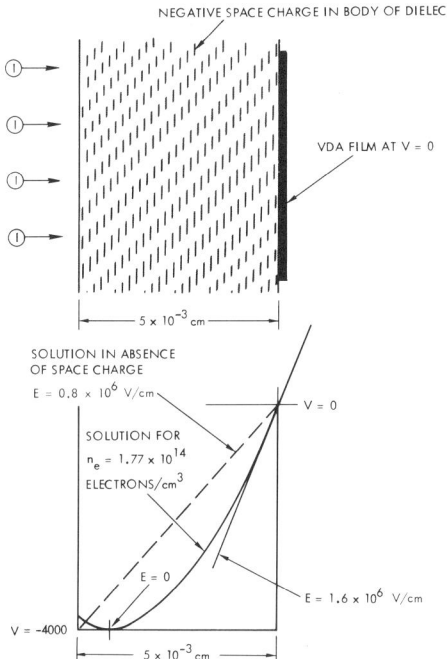

Fig. 11. Possible effects of space charge in alteration of electric field in dielectric material for incident electron beam. Front-to-back potential increment held fixed.

a front-to-back potential difference of 4×10^3 V would result purely from space-charge curvature of the potential. In this space-charge dominated example, E vanishes near the forward face of the dielectric and is 1.6 MV/cm at the rear face. However, this result is not consistent with the earlier measurements of total charge per area at the back face (integrated displacement current) and indicated front surface potential (varying applied potential to the grid). The likely situation is that space charge effects, from trapped charge carriers in the dielectrics, are present but do not dominate the evolution of the steady state electric field. Present experiments are not sufficient to define the extent of this trapped charge or the mobility and lifetime of the charge carriers.

5.4 Bulk Resistivity

The steady state drainage current for the 50 cm² Kapton sample is ~16 nA for a drainage current density of 3.2×10^{-10}

A/cm^2. From ΔV = 4000 V, and film thickness = 5 × 10^{-3} cm, a bulk resistivity of ρ = 2.5 × 10^{15} Ω-cm is derived at E = 8 × 10^5 V/cm.

The measured bulk resistivity of 2.5 × 10^{15} Ω-cm is lower, by a factor of 400, from the nominally quoted figure of 10^{18} Ω-cm for Kapton. This reduction is not surprising in view of the very high level of electric stress. Two other factors also may be considered as possible sources of reduced resistivity. The first of these is energy release in the deposition of the energetic primary electrons. These electrons impact on the surface, in a steady state, with an average energy of ~2000 eV. Although most of the energy loss in the stopping of the electron is local (in the ~1000 Å layer which marks depth of penetration), some of the stoppage results in the emission of short wavelength photons which can be absorbed in other regions of the dielectric. This bombardment induced conductivity would depend upon flux levels in the incident beam, and would cause the drainage current to be dependent upon both the energy and current density of the incoming primary electrons. A second possible mechanism causing conduction in this present "dark" experimental configuration (electrons present, uv absent) is photon emission from the electron gun filament. While the range of photon emission from the filament is predominantly at long wavelengths, some photoconduction in the dielectric may be present at other than uv wavelength.

5.5 Photoconduction for Applied uv

Figure 12 illustrates the drainage current through the Kapton for 6-keV incident electrons for uv-absent and uv-present. In the configuration used, the uv source-to-dielectric-

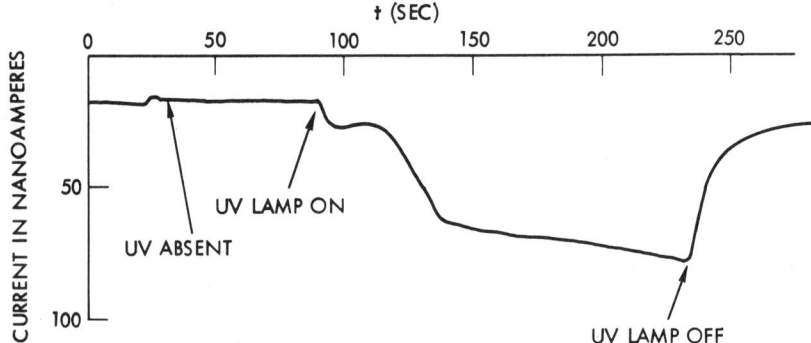

Fig. 12. Conduction current to aluminum film collector for uv-absent and uv-present for steady state 6-keV electron stream onto Kapton surface.

sample separation distance was 13 cm, so uv intensity was considerably less than "one sun," by any standard of equivalency. Nevertheless, the onset of uv caused a measurable jump in conducted current, and, as the uv source intensity rose (visibly increasing lamp brightness), photoconduction continued to rise reaching drainage current levels ∼4 times the "dark" current. Turning off the uv caused a gradual return to the former drainage current level.

Photoconduction in dielectrics will depend upon spectral distribution of the incident light, angle of incidence, and absolute intensity. The experiments with the Hg uv source clearly do not explore the photoconduction of Kapton under solar irradiation. The experiments do confirm, however, the importance of photoconduction effects, and indicate future directions for EST measurements programs.

6. Drainage Current Measurements as a Function of Electron Swarm Energy

The bulk resistivity as a function of electric stress is determined by varying the incident beam energy and the resultant front surface potential. An alternate description is to treat E (and also front-to-back potential) as unknown quantities, since in the equilibration in space neither E nor ΔV are determined, and simply to measure the bulk conduction current as a function of incident electron energy. Figure 13 illustrates this drainage current for incident electron energy in the range from 4 to 10 keV, and for both uv-absent and present. The principal feature of these data are the rapid rise of drainage current for increases in electron acceleration voltage. Inouye et al.[5] have reduced the conduction currents into a film resistance given by

$$R = 3.58(10)^{26} V^{-3.5} \ \Omega/cm^2 \tag{2}$$

for uv-absent and

$$R = 1.51(10)^{25} V^{-3.28} \ \Omega/cm^2 \tag{3}$$

for uv-present where V is electron acceleration potential in volts. These formulas are used to determine overall spacecraft electrical equilibration. The uv levels applied here have been discussed previously (Sec. 4.) and do not constitute a "one-sun" equivalency.

In the data of Fig. 13 and for 6.0-keV electrons, the drainage current recorded was 0.28 nA/cm². In the data of Sec. 5.4, this same beam energy resulted in a drainage of .32

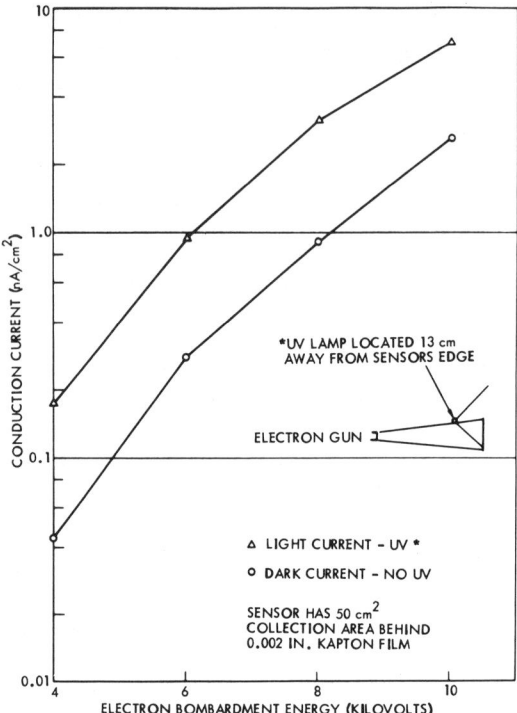

Fig. 13. Conduction current density through Kapton as a function of incident electron bombardment voltage for uv-absent and uv-present conditions.

nA/cm^2. The Kapton used is the same sample for both measurements, with the smaller drainage current density resulting from the first exposure. This suggests that high electric stress levels and the resultant conduction current cause permanent reductions in the bulk resistivity of the Kapton. At the conclusion of the exposures and utilizing a 2.0-keV electron beam plus a grid at -1.8 keV, in order to generate a front surface potential of -1170 V, a drainage current density of 9.2 pA/cm^2 was observed for a measured resistivity of 2.5×10^{16} Ω-cm at E = 0.234 MV/cm. The "low" value of resistivity at this stress level appears to be degraded significantly from the resistivity that would have been present on the unstressed sample.

7. Metal-to-Metal Arc Analysis

In order to discuss and evaluate the metal-to-metal arc discharge data presented in the earlier sections a conceptual arc model will be developed. The samples used in the series of metal-to-metal arc tests had a total metallic surface area of

~1000 cm^2, and front-to-back surface capacitance of ~50 nF. Stored energy in this front-to-back surface metal-to-metal capacitance is ~3 joules at 10,000 V. When surface breakdown of the dielectric initiates, transfer of this stored energy from the capacitance to the arc takes place through the VDA film.

7.1 Transfer of Stored Energy to the Region of Breakdown

The properties of the VDA film, the intervening dielectric, and the rear surface metal are essentially those of a parallel bar (or "slab") transmission line.[10] Figure 14 illustrates such a line denoting its elements and its connection to the discharge impedance. If z denotes distance along the metal

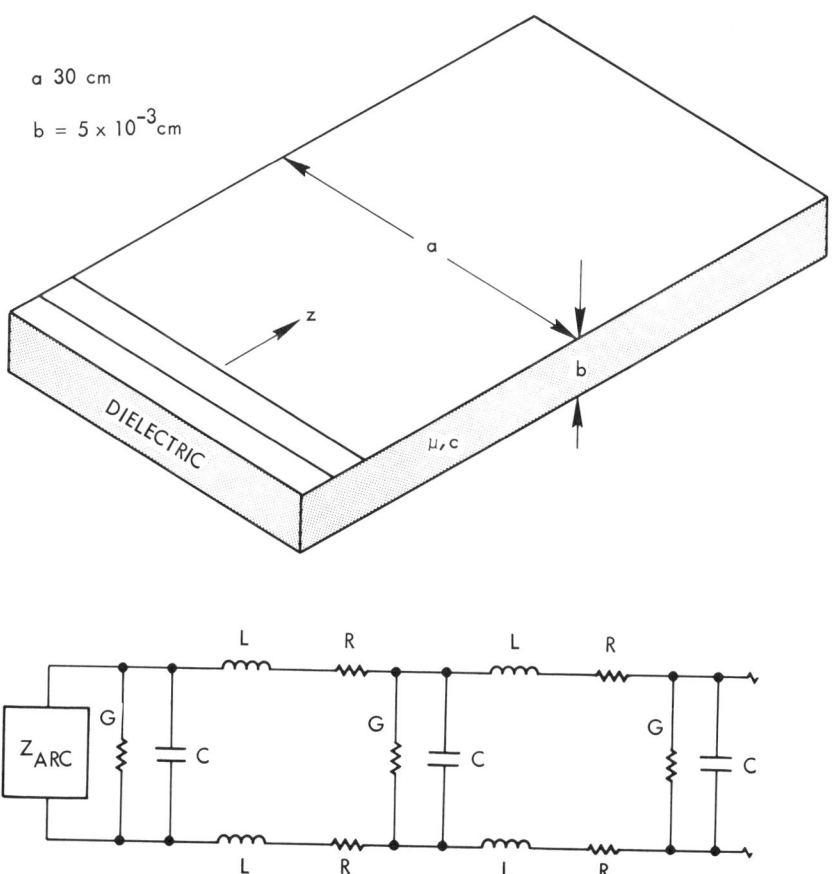

Fig. 14. Representation of VDA/Kapton/metal sample as a "slab" transmission line with metal-to-metal arc as load (Z_{ARC}).

face moving away from the dielectric gap, the equations for potential and current along the transmission line are

$$\partial V/\partial z = -L(\partial I/\partial t) - RI \tag{4}$$

and

$$\partial I/\partial z = -C(\partial V/\partial t) - GV \tag{5}$$

where L and C are inductance and capacitance, R is series resistance, and G is shunt capacitance per unit length of the line. Using values of μ and ε characteristic of the Kapton dielectric, and the dimensions of the sample (30 cm width, front-to-back separation of 5×10^{-3} cm), $L = 2 \times 10^{-10}$ H/m and $C = 2 \times 10^{-7}$ F/m. The VDA film (for initial thickness of 1000 Å) has $R = 1\ \Omega/m$. Shunt conductance is more difficult to calculate, but using a range of conductivity for Kapton in this high stress condition, described earlier, yields values from 10^{-10} mho/m to 10^{-12} mho/m. The response of a slab transmission line of these parameters to a signal of frequency ω is a propagation constant, γ, given by

$$\gamma = [(R + j\omega L)(G + j\omega C)]^{1/2} \tag{6}$$

where

$$\gamma = \left[(1 + j\omega(2 \times 10^{-10}))(10^{-11} + j\omega(2 \times 10^{-7}))\right]^{1/2}. \tag{7}$$

Unless ω is extremely large ($\sim 10^{10}$ rad/sec) the $j\omega L$ term may be neglected, and unless ω is very small ($\sim 10^{-3}$ rad/sec) the shunt conductance may be neglected. For $10^{-3} < \omega < 10^{10}$ rad/sec then

$$\gamma = \left(j\omega(2 \times 10^{-7})\right)^{1/2} \tag{8}$$

which yields an attenuation constant α of

$$\alpha = (\omega \times 10^{-7})^{1/2} \tag{9}$$

with a phase constant β of

$$\beta = (\omega \times 10^{-7})^{1/2} \tag{10}$$

where both α and β are in inverse meters. For values of $\omega = 2\pi \times 10^7$ (f = 10 MHz), $\omega = \sqrt{2\pi}$, and ωz for $z = 0.3$ m (the length of the sample in the z direction) is ~ 0.8. There would be, thus, significant attenuation of 10 MHz signals moving along 30 cm of this sample face. This offers, in turn, a partial explanation of observed rates of rise in the arc current, noting that energy transfer from various portions of the sample to the discharge region cannot occur more rapidly than approximately 100 nsec, which is very nearly the observed period of growth of arc current. It should be emphasized that the transmission line aspects presented here provide only a partial explanation, since arc impedance itself during the discharge formation also may exercise significant control on the total discharge process.

The effectiveness of energy transfer of the arc will be determined primarily by resistance in the metal film. Thickening of the VDA layer will lower the value of R in Sec. 5.1, and extend the bandwidth of the transmission line so that more rapid energy transfer occurs. Previous measurements of VDA/ground strap current carrying properties[11] indicate the desirability of an increase in film thickness to prevent ground-strap removal as a result of current pulses. Viewing both those results and present arc analyses would tend to favor, if anything, increases in R (reduction in film thickness).

The rate of growth of the plasma "film" over the dielectric surface will depend upon the total voltage drop across the dielectric gap, gap width and substrate material. If the dielectric material is vaporized easily and the resulting atoms and molecules are ionized easily, plasma growth rate is enhanced, the arc impedance will diminish at a more rapid rate, with larger dI_A/dt and I_A. Surface films on the substrate with comparatively reduced bonding to the underlying dielectric, might provide for more rapid plasma growth than for "clean" dielectric surfaces. Even among clean dielectrics considerable variations may occur in ultimate arc resistance.

7.2 Arc Formation and Motion

For surface breakdown, the level of electric stress at which breakdown initiates is approximately 1000 V/mm. It does not seem likely that breakdown at this comparatively low stress level can be initiated by electron acceleration, collision, and secondary electron excitation from valence to conduction bands by electrons, which remain wholly within the dielectric material. Rather, it seems likely that electrons moving in the regions above (but very near) the surface, because of increased mean-free-path before collision, would be more capable of gaining energy from the electric field to that point where their impact with the material could give rise to electron excitation to conduction bands (or to the "near"-surface region occupied by the primary electron). The energetic "primary" electron impact also should release short wavelength photons, which can penetrate the dielectric and cause additional electron excitation to conduction bands.

The growth of an electron colony in the conduction band could be regarded as an additional step in a breakdown chain leading ultimately to an electron avalanche and high conductivity of current. However, for surface breakdown, a more reasonable process would involve the release of atoms from the dielectric surface and their subsequent ionization. This material removal and plasma growth certainly occurs in those regions of VDA where removal is evident and large emission levels of light indicate plasma formation. The discussion here will consider that material removal and ionization occurs for both the dielectric surface and the connecting regions of VDA, and that the bulk of conducted current is in this layer of plasma. Current flows very near the surface of the magnitude observed in the EST tests create significant levels of magnetic fields which, in turn, interact with the arc current producing the field. The result of this $\vec{I}_A \times \vec{B}$ interaction is an outward force on the principal charge carriers (the electrons), and they are accelerated away from the surface. Because of space charge neutralization requirements, however, the electrons cannot move without an accompanying motion of the ions, and thus both electrons and ions are accelerated away from the surface. Figure 15 illustrates this acceleration process.

The outward motion of the plasma and its conducted arc current increases the inductance of the current flow, and in

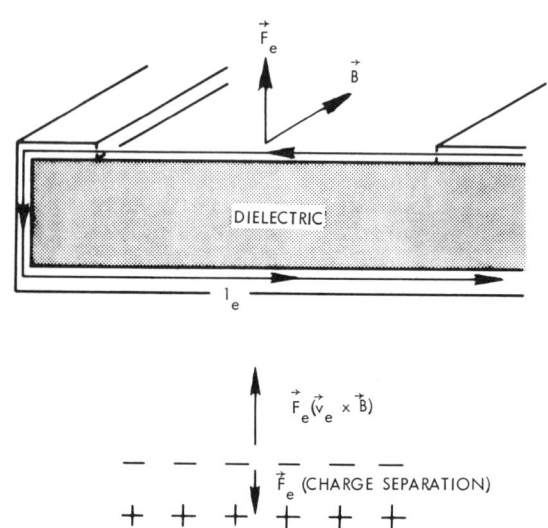

Fig. 15. Illustration of arc current flow, $\vec{B}$ field generation, $\vec{I} \times \vec{B}$ acceleration of arc electrons, and charge separation field causing resultant arc ion acceleration.

all likelihood, increases the resistance of the arc. Under this process of arc motion and disruption (and more distant placement from the surface where it ingests and ionizes material), an eventual decay of arc current may be formulated. A limitation in arc current emerges, then, from this arc acceleration process. Although the present arc model is largely conceptual, some indications of arc motion may be gained from simple analysis. If the arc current is I_A, the length of arc is ℓ and the magnetic field created by current flow is B, then the total force on the arc is given by

$$F_T = I_A B \ell \tag{11}$$

where F is in newtons for I_A in amperes, B in webers per square meter, and ℓ in meters. For the geometry illustrated in Fig. 15, the magnetic field will be given approximately by

$$B = \mu_0 I_A / 2\pi x \tag{12}$$

where $\mu_0 = 4\pi \times 10^{-7}$ and x is current flow separation distance in meters. The total force on the arc causes plasma acceleration given by

$$d^2x/dt^2 = F_T/M_T \tag{13}$$

where M_T is the total mass of material in the arc. Since the mass of the plasma is principally ions, the total mass of the plasma is

$$M_T = n_+ A_A \ell M_+ \tag{14}$$

where n_+ is ion density per cubic meter, A_A is arc area in square meters, ℓ is arc length, and M_+ is average ion mass in kilograms. A final equation for use in the solution of Eq. (13) is

$$I_A = n_e e A_A v_e \tag{15}$$

where n_e is electron density per cubic meter, e is electron charge in coulombs, A_A is arc area and v_e is average electron velocity in meters per second. Inserting Eqs. (11, 12, 14, and 15) into Eq. (13) yields

$$d^2x/dt^2 = \mu_0 n_e e v_e I_A / 2\pi x n_+ M_+ \tag{16}$$

and, since $n_e = n_+$ from space charge neutrality,

$$d^2x/dt^2 = \mu_0 e v_e I_A / 2\pi x M_+ \tag{17}$$

Denoting $\eta = x/x_0$, where x_0 is initially current-to-current separation, and noting that $(d\eta/dt) = 0$ at $t = 0$, leads to a solution of Eq. (17) in terms of a relative growth of the separation distance, γ, where

$$\int_1^\gamma d\eta/\sqrt{\ln\eta} = \sqrt{2K}\, t \qquad (18)$$

where

$$K = \mu_0 e v_e I_A / 2\pi x_0^2 M_+$$

is in inverse seconds squared and t is in seconds. Figure 16 provides an integrated left-hand side of Eq. (18). It may be seen from Fig. 16 that substantial reorientation occurs in the current-to-current separation for $\sqrt{2K}\, t \sim 2$ or $t \sim (s/K)^{1/2}$. Inserting values into K, and using $I_A = 1000$ A, $v_e = 10^6$ m/sec, $M_+ = 2 \times 10^{-26}$ kg, and $x_0 = 10^{-3}$ m, leads to $K = 2 \times 10^{15}$ and $(2/K)^{1/2} = 30$ nsec. If x_0 is reduced to 10^{-4} m, the same proportionate change will occur in the relative x in ~ 3 nsec. There is, thus, appreciable motion of an arc under such accelerative forces, and the decay of arc current may result from this upward acceleration, first as it increases the current return loop area and loop inductance, and second, as it removes the arc from the surface and the material which it ingests and ionizes.

The properties of breakdowns in the bulk of the dielectric may be expected to differ significantly from those on the sur-

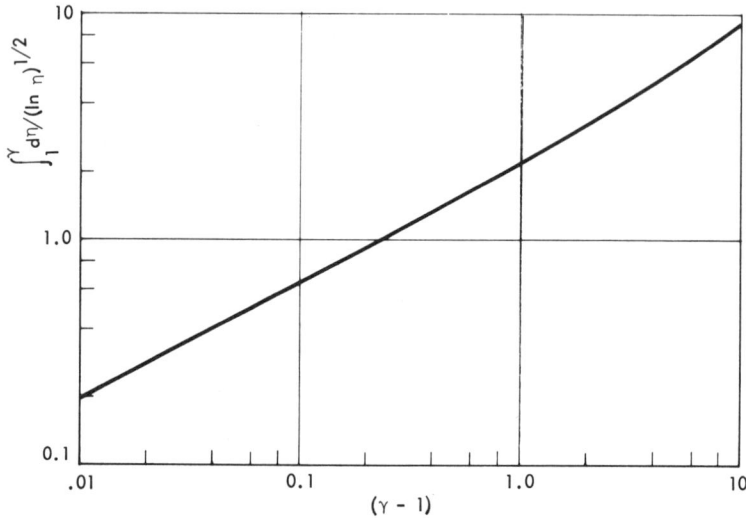

Fig. 16. Value of $\int_1^\gamma d\eta/(\ln\eta)^{1/2}$ of a function of integral upper end point, γ.

face. In a bulk breakdown, initial stress levels are much higher (of the order of 10^6 V/cm) and electron avalanche growth must take place totally within the material, rather than both in and near the material, as in surface breakdown. Disruption of the material and plasma formation occurs, however, for both cases of breakdown. The arc length in the bulk breakdown may be much shorter than for surface discharges (present Kapton samples are 5×10^{-3} cm-thick), and arc motion in current generated B fields cannot be away from the surface for at least those portions of the arc penetrating the bulk material. This could prevent (or delay) those arc motion effects limiting surface discharge growth and duration.

For bulk breakdown in an interior portion of a dielectric sheet and from a forward to a rear conducting face, current flow results in considerably less B field generation and the resulting arc is essentially force free. There would be, thus, less acceleration of the plasma away from material surfaces "feeding" the breakdown region and larger values of both I_A and dI_A/dt would be expected. However, the radiated impulse from these bulk breakdown arc current flows might be expected to be reduced because of symmetries in the current flow (and resulting cancellations in $\vec{E}$ and $\vec{H}$). Effects of current transients on spacecraft materials and spacecraft structures are under investigation[12,13] and should be extended to treat presently determined arc currents.

Another difference between surface and bulk breakdown is in energy transfer to the arc from storage in the "capacitor." Bulk breakdown, when it occurs away from an "edge" in the material, is fed essentially by a cylindrical rather than a linear transmission line. The bandwidth of this cylindrical geometry is expected to be larger than that of the linear line. The arc current growth rate, then, may be expected to reach higher levels in the bulk breakdown than for the surface charge. With increases in both I_A and dI_A/dt, arcs from bulk breakdown may exercise even greater impact on spacecraft operation than those from surface breakdowns. Both forms of discharges, however, have significantly large I_A and dI_A/dt, and a need for further experimental and analytical examination is indicated.

References

[1] DeForest, S. E., "Spacecraft Charging at Synchronous Orbit," Journal of Geophysical Research, Vol. 77, Feb. 1972, pp. 651-659.

[2] Rosen, A., Fredricks, R. W., Inouye, G. T., Sanders, N. L., Scarf, F. L., Greenstadt, E. W., Vogl, J. L., and Sellen Jr., J. M., "RGA Analysis; Findings Regarding Correlation of Spacecraft Anomalies with Magnetic Substorms and Laboratory Test Results," Rept. 09670-7020-R0-00, 1972, TRW Defense & Space Systems, Redondo Beach, Calif.

[3] Fredricks, R. W. and Scarf, F. L., "Observations of Spacecraft Charging Effects in Energetic Plasma Regions," Photon and Particle Interactions with Surfaces in Space, edited by R. J. L. Grard, D. Reidel Publishing Co., Dodrecht, Holland, 1973.

[4] "TDAL (Tunnel Diode Amplifier-Limiters) Gain State Analysis - Final Report," Rept. 09670-7040-RU-00, Oct. 18, 1973, TRW Defense & Space Systems, Redondo Beach, Calif.

[5] Inouye, G. T., Vogl, J. L., Sellen Jr., J. M., DeForest, S. E., and Rosen, A., "Final Report Spacecraft Charging Analysis: A Study and Analysis of the DSCS II Spacecraft Orbital Charging Phenomena," TRW Defense & Space Systems, Redondo Beach, Calif., Aug. 16, 1974.

[6] Komatsu, G. K., Hoffmaster, D. K., Inouye, G. T., and Sellen Jr., J. M., "Observations of Behavior of Insulating and Electrically Isolated Conducting Materials in a High Energy Electron Swarm Tunnel," Rept. 4360.4.2.73-99, Oct. 2, 1973, TRW Defense & Space Systems, Redondo Beach, Calif.

[7] Sellen Jr., J. M. "Electrical Equilibration of Conducting and Insulating Materials in the Presence of Energetic Electrons and Ultraviolet Light," Rept. 4351.3.74-44, Oct. 31, 1974, TRW Defense & Space Systems, Redondo Beach, Calif.

[8] Balmain, K. G., Orszag, M. and Kremer, P., "Surface Discharge on Spacecraft Dielectrics in a Scanning Electron Microscope," Paper SA69, Spring Annual Meeting of the American Geophysical Union, Washington D.C., June 16-19, 1975; presented elsewhere in this volume.

[9] Adamo, R. C. and Nanevicz, J. E., "Spacecraft-Charging Studies of Voltage Breakdown Processes on Spacecraft Thermal Control Mirrors," Paper SA42, Spring Annual Meeting of the American Geophysical Union, Washington D.C., June 16-19, 1975; presented elsewhere in this volume.

SURFACE DISCHARGES ON SPACECRAFT DIELECTRICS
IN A SCANNING ELECTRON MICROSCOPE

K.G.Balmain,[*] M.Orszag,[†] and P.Kremer[‡]

Department of Electrical Engineering
University of Toronto, Toronto, Canada

Abstract

The scanning electron microscope (SEM) is shown to be a useful device for the study of surface electrical discharges on spacecraft materials. Discharges were observed on specimens of Kapton, Kapton-fiber glass laminate, Kevlar-fiber glass laminate, and Teflon, the observations involving primarily measurements of secondary-electron emission and specimen-current frequency spectrum. The spectra of all these materials were similar, with a constant amplitude of -45 to -50 dBm at 300 kHz bandwidth, flat up to 100 MHz, and dropping off above that. With Teflon, however, the frequency of discharge occurrence was several times that of the other materials. Discontinuities in secondary-electron SEM photographs suggest discharge "cleanoff" of a large fraction of the deposited charge, and a mechanism of discharge propagation through the submerged charge layer is proposed to explain this effect.

Presented in part as Paper SA69 at the Spring Annual Meeting of the AGU (cosponsored by AIAA), Washington D.C., June 16-19, 1975. The research was supported by the Communications Research Centre, Department of Communications, Canada, under Department of Supply and Services Contract No.DSS 36100-4-0235. Permission by the Department of Communications to publish the results of this study is acknowledged. Support was also provided by the National Research Council of Canada, under Grant No. A-4140. Discussions with R.E.Barrington are gratefully acknowledged.
*Professor of Electrical Engineering.
†Post-doctoral Assistant.
‡Engineering Technologist.

Introduction

There is now little doubt that a satellite in synchronous orbit can accumulate electric charge in large quantities, in the presence of the high-energy electrons associated with magnetic storms and in the absence of sufficient charge drainage by mechanisms such as photoemission. Because the exposed surface of a typical satellite is composed largely of insulating materials (dielectrics), and because the charge buildup and drainage effects vary over the surface, the whole-satellite charging, already mentioned, must be accompanied by strong differential charging. Subsequent local discharge by electrical breakdowns and arcs is clearly a possibility and could have led to the satellite operational anomalies which have been observed. Synchronous-orbit observations and also some relevant laboratory experiments have been reviewed thoroughly by Rosen[1], making unnecessary any further general remarks here.

Let us turn now to the specifics of laboratory simulation of spacecraft-charging conditions. The necessity for simulation in a vacuum chamber is a crucial limitation because the larger the chamber, the better the simulation ought to be. On the other hand, the smaller the chamber, the more inexpensive the experiments become, and the easier it is to make changes. A small, highly functional, experimental vacuum chamber complete with electron-beam source is provided by a typical scanning electron microscope (SEM), whose normal mode of operation involves measuring secondary-electron emission and using this measurement to create an image by synchronization with the incident beam scan. Such secondary-electron images have been used by Balmain[2] to display charge-accumulation patterns on spacecraft dielectrics and to provide tentative evidence for the existence of impulsive discharge events on Kapton and Kapton-fiber glass laminate, events indicated by momentary increases in secondary-electron emission. The further development of SEM simulation of spacecraft dielectric discharges is described in the remainder of this paper.

Experiments

The Scanning Electron Microscope

Each test specimen of spacecraft dielectric material was mounted in the SEM vacuum chamber as shown in Fig. 1. Experiments involved measurement of secondary electrons and measurement of current passing through the specimen to the copper plate beneath it. For secondary-electron imaging, the signal

SURFACE DISCHARGES ON SPACECRAFT DIELECTRICS 215

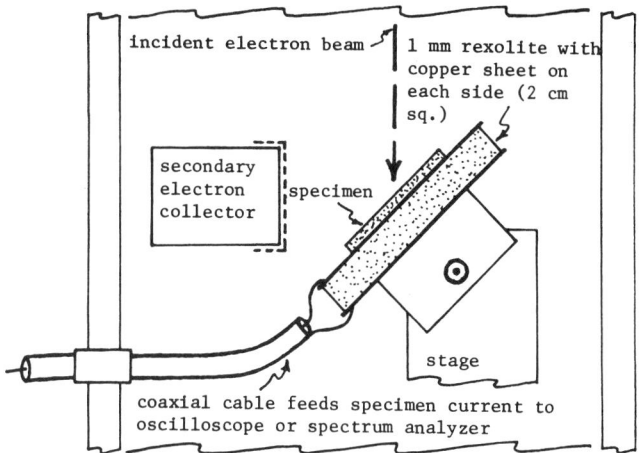

Fig. 1 Arrangement of components in specimen vacuum chamber of scanning electron microscope.

from the electron collector intensity-modulates a cathode-ray tube (CRT) whose beam position is synchronized with the position on the specimen of the SEM incident electron beam, which is scanned in a left-to-right raster going from top to bottom (with reference to the CRT display). Consequently, the resulting photographs must be thought of not only as recordings in space but also as recordings in time.

Secondary-Emission Photography

One effect of negative-charge accumulation on a small specimen is to deflect the incident electron beam toward the edge of the specimen. However, the SEM image records the intended, rather than the actual point of contact of the beam with the specimen. Therefore, on a negatively charged specimen, two identifiable points on either side of the center of the specimen will appear to be closer together than is actually the case, and any decrease in the amount of charge will cause the apparent separation between the two points to increase. This is illustrated in Fig. 2, in which two straight scratches on a Kapton sample suddenly appear to jump apart. This discontinuity in the photograph is accompanied by a white streak, indicating a sharp increase in electron emission from the specimen. The increase in electron emission clearly is consistent with the reduction in surface charge, so it is definite that we can think in terms of "discharge events" occurring on the specimen. The locations on the specimen of white streaks, which are actually events in time rather than space. Furthermore, the discharges need not occur at a

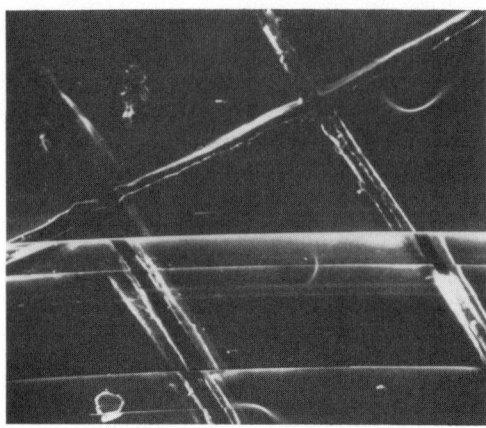

Fig. 2 Secondary-electron photograph of Kapton at 20-kV beam voltage and x 280, showing discharge events.

well-defined point but in some sense could be distributed over the specimen surface. Such a view is supported by the observation that many of the discharge events restore the apparent scratch separation essentially to the uncharged separation, indicating "cleanoff" of a large fraction of the accumulated charge at each event.

The discharge events just described were observed on Kapton, Kapton-fiber glass laminate, and Kevlar-fiber glass laminate. The appearance of the events in the photographs ranged from a few heavy streaks, as in Fig. 2, to a "snowstorm" of very short streaks. Discharge events were never observed on typical solar cell cover glass (5% CeO-doped, with MgO coating), on which the short decay time of accumulated charge (a few seconds) apparently did not permit the buildup of strong differential charge in the SEM environment, even with a beam accelerating voltage of 30 kV.

Specimen Current

The specimen current was measured first by connecting the stage (on which the dielectric was mounted) to the "specimen current amplifier" on the SEM and recording the result. A typical result for Kapton is shown in Fig. 3 along with a corresponding measurement of secondary emission. In this case, the specimen current has a background value of about 10^{-10} A caused mainly by displacement current accompanying the slow temporal buildup of electric charge on the surface of the di-

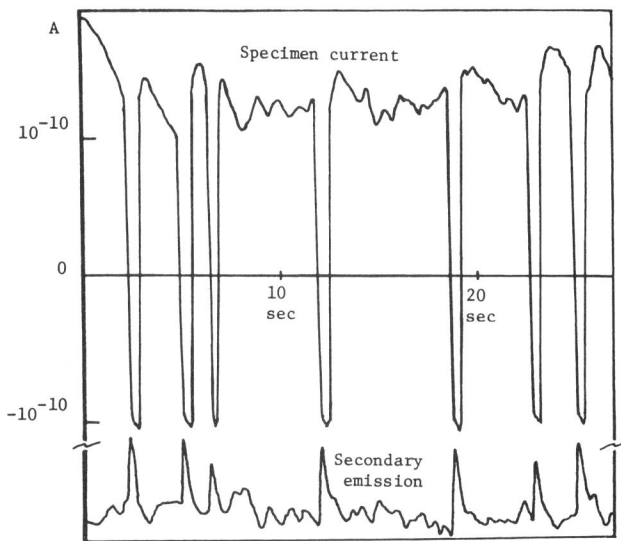

Fig. 3 Discharge events indicated by specimen current and secondary emission, for Kapton 0.1 mm thick and 20-kV beam voltage.

electric. From time to time there is a discharge event producing the burst of secondaries already discussed. Simultaneously, there is a reversal of specimen current, driving the amplifier to saturation in the negative direction, which is the direction of electrons flowing upward toward the specimen surface, or positive charges flowing downward. Such an effect could be entirely capacitive, the upward release of specimen surface electrons being accompanied by the downward release of positive charge previously attracted to the specimen stage by the slow accumulation of charge on the specimen surface. Conduction through the dielectric can be ruled out as an explanation for the current pulses, because such conduction would have to involve the passage of surface electrons downward through the specimen.

To get a better idea of the rise and fall times of the specimen current pulses, the experiment was repeated using a 25-MHz storage oscilloscope. The negative pulses were observed as before, rising to several volts for a 1-MΩ oscilloscope input resistance. The fall time of the pulse was determined by the time constant of the oscilloscope input resistance and the total capacitance (oscilloscope + cable + specimen stage). However, the rise time was "instantaneous", independent of the time constant, as one might expect from the "instantaneous" release of charge from the specimen stage. As measured in a

250-MHz oscilloscope, the "instantaneous" rise times turned out to be between 4 and 8 nsec, as did the fall times when using a 50-Ω load. On rare occasions, strong positive pulses were observed with the 25-MHz oscilloscope, suggesting breakdown through the 125-μm thickness of the Kapton specimen. However, the vast majority of discharge events appear not to penetrate the specimen but rather to occur at or close to its surface.

Spectrum Analyzer Measurements

The specimen current discharge pulses were measured first by setting the spectrum analyzer at various fixed frequencies and, at each one, taking note of the amplitude of the strongest pulse in a 1-min interval, using a 10 kHz bandwidth. For this experiment, the SEM was not scanned but rather the spot beam was moved slowly back and forth by hand-adjustment of the deflection controls. This procedure gave many times more discharges per unit time than did the rapid beam scan used previously.

The results for Kapton and Kevlar-fiber glass laminate are shown in Fig. 4. The two spectra are essentially the same: flat up to 100 MHz followed by a dropoff of between 30 and 40 dB/decade of frequency. Such a spectrum with the 40 dB/dec-

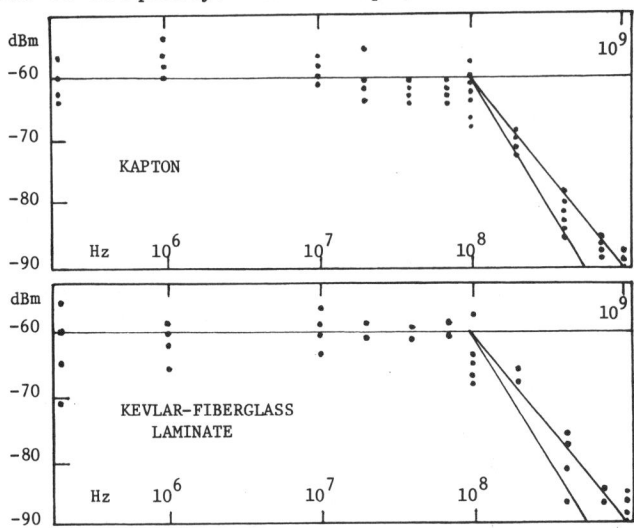

Fig. 4 Discharge spectra for specimens with 1 cm^2 area. Each dot is amplitude of strongest discharge pulse in 1 min interval. Lines indicate dropoff rates of 30 and 40 dB/decade of frequency. An intermediate frequency bandwidth of 10 kHz was used, with a 10-kHz low-pass video filter.

ade dropoff figure is characteristic of continuous-function pulses with rise and fall times of the order of a few nanoseconds.[3,4] Sharper pulses involving step-discontinuities produce slower dropoff rates.

A second type of spectrum analysis experiment involved using a wider bandwidth of 300 kHz and sweeping the frequency slowly from 100-1100 MHz while recording the detected discharge pulses on a storage oscilloscope. The materials tested were Kapton, Kevlar-fiber glass laminate, and Teflon. The spectral dropoff rates were similar in all cases: 30-40 dB/decade from 200-500 MHz and a somewhat slower, more irregular dropoff up to 1100 MHz. The results for the Kevlar-fiber glass laminate were indistinguishable from those for the Kapton. The Teflon, however, exhibited a frequency of discharge occurrence 5-6 times higher than the Kapton while giving pulse amplitudes 15 dB lower than the Kapton.

It should be mentioned that the SEM beam accelerating voltage threshold for the observation of discharge events was normally 16-18 kV for all materials, with essentially no change in measured spectrum from 20 to 30 kV. The addition of a large amount of contaminant to the dielectric surface, such as a smear of diffusion pump oil, merely raised the discharge threshold to between 20 and 25 kV. Variation in specimen surface area from 1 mm^2 to 250 times that had no effect on the spectrum in this type of measurement. Also, specimen thickness variations from 0.1 to 0.5 mm had no effect. In some cases, however, beam incidence on the edge of the specimen produced increased frequency of discharge occurrence. In addition, the light output from the discharges was photographed, the results indicating that the brightest part of each discharge is highly localized at the point of beam incidence. An electrically floating vacuum-deposited layer of gold, about 200-Å- thick, completely stopped the discharges in a sequence of experiments on Teflon.

In order to determine the effect of incident-beam current density, the beam was defocused by varying the "working distance" as measured below the final aperture of the instrument. A working distance set to give focus 15 mm above the specimen produces a beam diameter somewhat greater than 1 mm and a current density of the order of 10 nA/cm^2 (as expected in synchronous orbit under storm conditions); whereas a working distance adjusted close to focus on the specimen surface produces a current density of the order of 10^4 times as great (all of the experiments described heretofore employed a focused electron beam). Some of the results for Teflon are

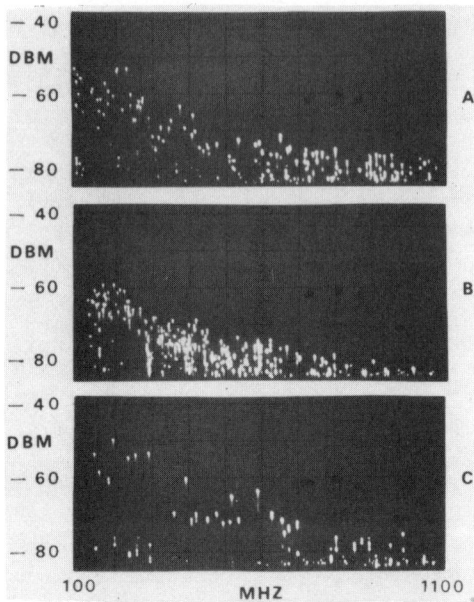

Fig. 5 Spectrum of Teflon discharges at 20 kV, for a 200-sec time interval and 300-kHz-IF bandwidth. Location of beam focus relative to specimen: (a) 25 mm below, (b) at surface, and (c) 15 mm above (effective beam diam ~ 1 mm).

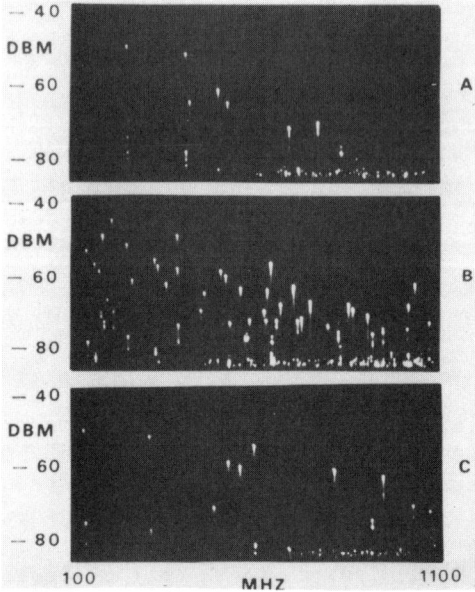

Fig. 6 Spectrum of Kapton discharges at 30 kV, for a 400-sec time interval and 300-kHz-IF bandwidth. Location of beam focus relative to specimen: (a) 15 mm below, (b) at surface, and (c) 11 mm above (effective beam diameter ~ 1 mm).

shown in Fig. 5, from which it is clear that a focused beam produces 3-4 times the frequency of discharge occurrence but 10-dB lower amplitude, compared to the defocused beam. For Kapton, as shown in Fig. 6, the effect on frequency of occurrence is the same, but the spectrum amplitude is unaffected. This latter point is especially worthy of note because it suggests that spectrum amplitude measurements on Kapton in an SEM are directly applicable to large-scale situations with broad incident beams. Applicability of the Teflon results is much less certain, but an approximate procedure would be simply to add 10 dB to the spectrum amplitudes obtained with a focused beam. It is interesting to note that such a 10-dB-adjustment brings the flat part of the Teflon spectral amplitude up to -50 dBm, fairly close to the Kapton level of -45 dBm.

Interpretation

For each electron incident on a sample of spacecraft dielectric, a net negative charge is deposited because, on the average, less than one secondary electron is emitted. This occurs provided that the incident-beam energy is above a threshold value (the "second crossover" on the yield vs energy curve). This threshold value is about 2000 eV for Teflon and 500 eV for Kapton, as reported by Willis and Skinner.[5]

High-energy (2 MeV) experiments have been reported by Gross on borosilicate glass[6] and Plexiglas[7], showing that the charge accumulates at a well-defined depth, and if the dielectric surface is pressed with a grounded needle, a discharge results. The discharge damage reveals that the path of the discharge is first "vertical" from the needle to the charge layer, and then "horizontal" through the charge layer in a feathery pattern (Lichtenberg figure). A lower-energy version of such a discharge could be triggered by the SEM electron beam, causing local breakdown between the submerged negative-charge layer and the dielectric surface (which would be slightly positive because of secondary emission). This initial breakdown, which has been discussed by Meulenberg,[8] then could be followed by a propagating discharge through the submerged charge layer. These ideas are illustrated in Fig. 7.

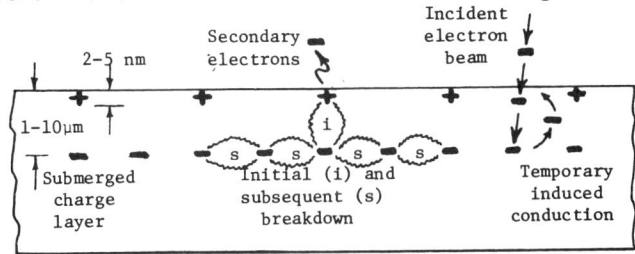

Fig. 7 Charging and discharging in a cross section of a dielectric specimen: illustration of some possible mechanisms.

If this discharge mechanism were operative, a thin layer of surface metallization would not stop the discharges. However, such a deduction ignores the phenomenon of beam-induced temporary conductivity (both "prompt" and "delayed" components)[9], which could drain at least part of the submerged charge to the surface metallization. Certainly much more experimentation needs to be done to establish clearly the effects of surface metallization, partly because these effects relate to a possible cure for the discharge problem, and partly because they help to reveal the nature of the discharge mechanism.

Conclusions

Surface discharges on thin specimens of spacecraft dielectric have been observed in the SEM by measuring secondary-electron emission and specimen current. The discharge currents have rise times of a few nanoseconds and fall times apparently determined by the external circuitry. The corresponding frequency spectra have a "knee" at about 100 MHz and are remarkably similar in amplitude and shape for different materials such as Kapton, Kapton-fiber glass laminate, Kevlar-fiber glass laminate, and Teflon (the Teflon amplitudes are about 5 dB lower than the other materials). However, Teflon exhibits a frequency of discharge occurrence which is several times higher than the other materials.

The discharge spectra (specimen-current measurements) have maximum levels from -45 dBm to -50 dBm for a 50-Ω input impedance and a 300-kHz bandwidth, strong enough to be classified definitely as sources of interference. It should be emphasized that these discharges are confined near the dielectric surface and are not dielectric-to-metal or metal-to-metal arcs. It is not clear yet how highly localized the surface discharges are, but there is some evidence for charge "cleanoff" over an appreciable area, possible via discharge propagation through the submerged layer of accumulated charge. No optically-visible discharge tracks have been observed yet, but SEM examination of much-used specimens reveals regular patterns of deep-seated alteration that could be either structural damage or long-lived charge accumulation.

References

[1] Rosen, A., "Large Discharges and Arcs on Spacecraft," *Astronautics & Aeronautics*, Vol.13, June 1975, pp. 36-44.

[2] Balmain, K.G., "Charging of Spacecraft Materials Simulated in a Scanning Electron Microscope", *Electronics Letters*, Vol.9, Nov.15, 1973, pp. 544-546.

[3] *Reference Data for Radio Engineers*, H.W.Sams & Co. Inc., Indianapolis, 1968, pp. 42.5-42.7.

[4] Bracewell, R., *The Fourier Transform and Its Applications*, McGraw-Hill, New York, 1965.

[5] Willis, R.F., and Skinner, D.K., "Secondary Electron Emission Yield Behavior of Polymers," *Solid State Communications*, Vol. 13, No. 6, 1973, pp. 685-688.

[6] Gross, B., "Irradiation Effects in Borosilicate Glass," *The Physical Review*, Vol.107, July 15, 1957, pp. 368-373.

[7] Gross, B., "Irradiation Effects in Plexiglas", *Journal of Polymer Science*, Vol. 27, 1958, pp. 135-143.

[8] Meulenberg, A., "Evidence for a New Discharge Mechanism for Dielectrics in a Plasma," *Spring Annual Meeting of the American Geophysical Union*, Paper SA70, Washington, D.C., June 16-19, 1975.

[9] Gross, B., Sessler, G.M., and West, J.E., "Charge Diagnostics for Electron-Irradiated Polymer Foils," *Applied Physics Letters* Vol. 22, April 1, 1973, pp. 315-316.

SPACECRAFT-CHARGING STUDIES OF VOLTAGE BREAKDOWN PROCESSES ON SPACECRAFT THERMAL CONTROL MIRRORS

R. C. Adamo[*] and J. E. Nanevicz[†]

Stanford Research Institute, Menlo Park, Calif.

Abstract

Data from experiments on a synchronous-orbit satellite indicated that electromagnetic noise pulses (attributed to voltage breakdowns on insulating surfaces) occurred on the satellite, particularly during magnetic substorm conditions when charging currents may be high. A laboratory experiment was conducted to simulate the charging of second-surface thermal-control mirrors and to investigate the electrical characteristics of the resulting voltage breakdowns. An electron gun was used to illuminate an area of approximately 500 cm^2 with an electron beam density of up to 100 nA/cm^2 at energies controllable from a few hundred electron volts to 20 keV. It was found that negative charge was deposited on the mirror surfaces and that at beam energies above 5 keV discharges occurred from the mirror surfaces. The data indicate that, during a discharge, negatively charged material is driven away from the surface to a distance of over 7.5 cm. Pulse rise times were roughly 0.02 μsec, and pulse durations of 1 μsec and more were measured. These parameters are of interest both in the calculation of noise signal characteristics and in providing insight into the breakdown mechanisms.

Presented as Paper SA42 at the American Geophysical Union Spring Annual Meeting, Washington, D. C., June 16-19, 1975.

[*]Research Engineer.
[†]Program Manager.

Introduction

In connection with the study of processes affecting the lifetime of a synchronous-orbit satellite, a set of instruments, designed for the investigation of electrostatic charging of rockets,[1] was repackaged and flown on a non-NASA synchronous-orbit satellite.[2,3] When the instrumentation was being planned, great importance was attached to the surface potential of the thermal-control mirrors on the satellite, since charge residing on these surfaces might accelerate the deposition of contaminants, which would affect thermal-control characteristics. Furthermore, if the potential were sufficiently high, it could lead to voltage breakdowns capable of producing electrical noise, of liberating contaminating materials and even of causing damage to the mirrors themselves.

For the satellite experiment, an electric-field meter was mounted in a panel of second-surface quartz mirrors used for thermal control of the satellite. It was desired that the system be capable of measuring high fields corresponding to voltage breakdown potentials, and at the same time be able to detect much lower potentials that still could deflect charged contamination particles. Since it was not clear what types of breakdown processes would occur under space conditions, it was felt that a laboratory simulation was the most straightforward scheme for obtaining the required information. It was observed also that the laboratory simulation would afford, in addition, an opportunity to study the breakdown processes associated with mirror-surface charging. Such information was needed for the interpretation of the outputs of the pulse counter system flown on the satellite.[2,3]

Laboratory Instruments and Results

The laboratory setup used for these experiments is shown in Fig. 1. A panel of 10 mil thick second-surface quartz thermal-control mirrors was mounted on insulating standoffs on the base of the vacuum system. The metal substrate of the mirror panel was returned to ground through an HP-425A current meter. An electron gun at the top of the chamber was used to illuminate the panel with a beam of 10-keV electrons.

STUDIES OF VOLTAGE BREAKDOWN PROCESSES

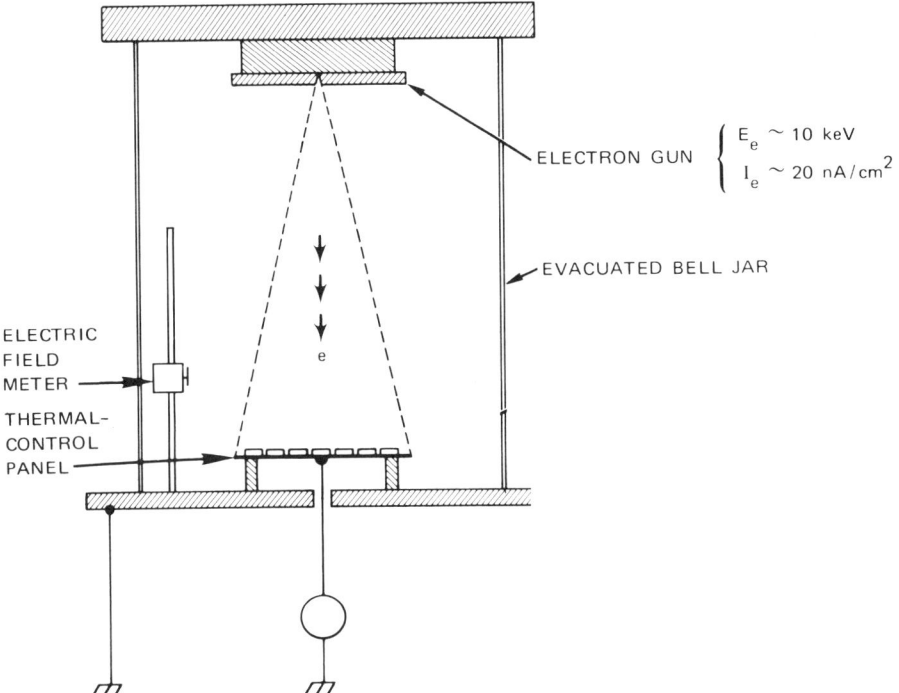

Fig. 1 Experimental setup for electron beam charging tests.

Provisions were made to vary the current density at the panel location from zero up to a maximum of 20 nA/cm^2. This density is approximately twice the typical average electron injection current densities reported for magnetic substorm conditions on satellites in synchronous orbit.[4,5]

An electric-field meter of the type flown on the satellite[2,3] was positioned at the side of the vacuum chamber. The field meter, mounted in an aluminum sheet, measured the field produced at its location by the charge residing on the surface of the mirror panels. The relationship between mirror surface potential and field meter output was established by placing a thin metal sheet over the mirrors, applying a known voltage to the sheet, and observing the resulting field meter output. It was recognized at the outset that the field meter would provide only an averaged reading for the potential of the mirror surfaces. Variations in potential from mirror to mirror could not be discerned. Since one of the main concerns in devising the voltage measuring system was that it should

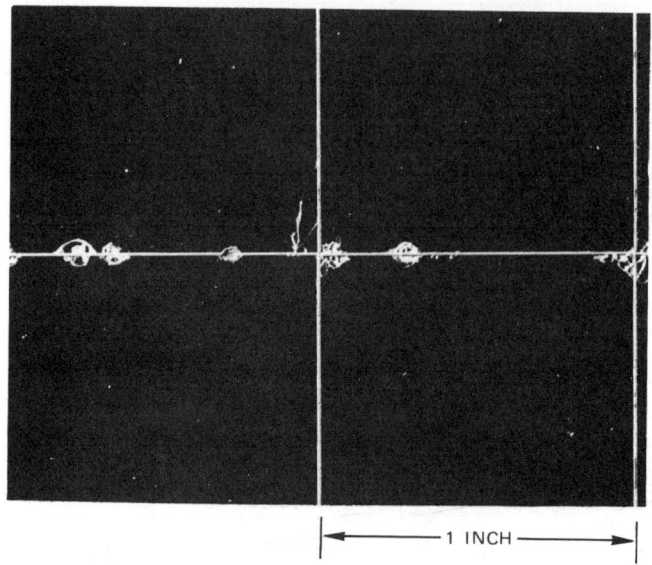

Fig. 2 Damage to quartz-mirrored thermal-control surfaces produced by electron-beam charging.

not influence the charging process, this minor shortcoming of the system was felt to be acceptable.

When the electron beam was turned on, periodic flashes of light were visually observed on the mirror panel. The rate of occurrence was proportional to the charging current density. Virtually all of the flashes occurred at the RTV filled gaps between the 1-in.^{2} mirrors. In addition to the periodic light flashes, a fluctuating blue glow pattern covered the mirror surfaces. The flashes occurred both as single events and as breakdowns that appeared to occur at the same time at two or more places.

Following irradiation by the electron beam, the mirror panel exhibited damage of the sort shown in Fig. 2. It was evident from careful inspection that, in addition to the erosion of the quartz illustrated in Fig. 2, RTV had been evolved from the joints and redeposited in the near vicinity. Only visual and microscopic examination of the mirrors was made on the present program. No measurements were made to estimate the degree of deterioration of thermal-control properties following irradiation by the electron beam.

Electrical measurements made in the setup of Fig. 1 produced data of the general character illustrated in Fig. 3. Here the panel was irradiated with a 10-keV beam of electrons with a current density of 1 nA/cm^2. When the beam was turned on, the electric field increased until it reached a level corresponding to a surface potential of 2 to 3 kV. At this time a breakdown indicated by a visible flash occurred, and the surface potential dropped to less than one-third of its value prior to the breakdown. It also should be noted that the occurrence of the breakdown increased the target current by roughly a factor of 3. This electrical behavior indicates that the observed breakdowns discharged a substantial portion of the mirror surface, since otherwise the surface potential reading would not be affected so strongly.

Figure 4 illustrates the behavior observed when the electron-beam current density was increased by a factor of roughly 3. Again, discharges occurred when the surface potential reached 2 to 3 kV. Each discharge evidently removes charge from a substantial fraction of the surface because the surface potential indication is strongly affected, as is the

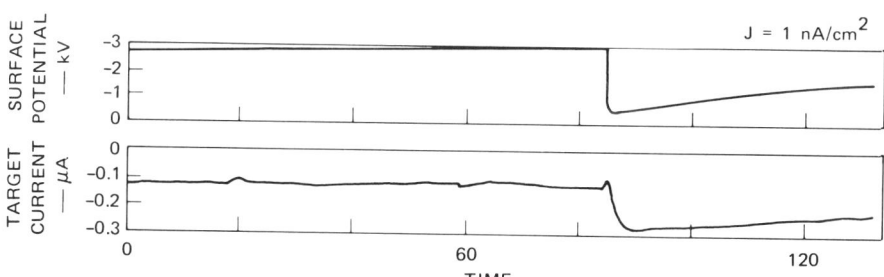

Fig. 3 Mirror surface potential under electron charging.

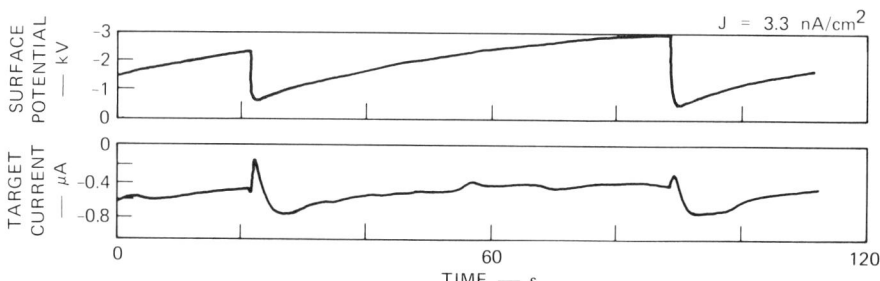

Fig. 4 Mirror-surface potential under electron charging with somewhat higher current.

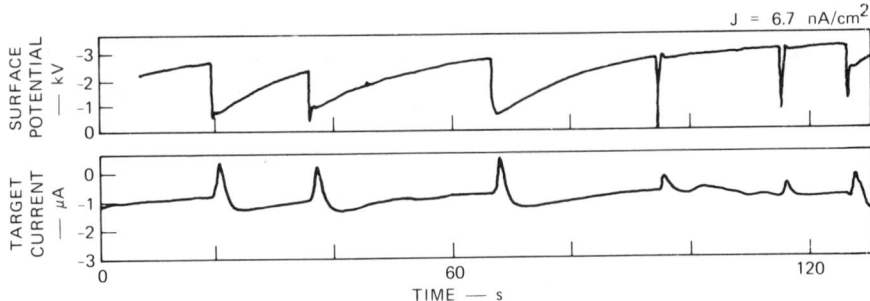

Fig. 5 Mirror-surface potential under electron charging with substantially higher current.

target current. It is evident that increasing the beam current density increases the repetition rate of the breakdown pulses (as one would expect in a relaxation oscillator).

A further increase in beam current produced records of the sort shown in Fig. 5. Here the repetition rate of the breakdowns is still higher. In the first three breakdowns shown on the record, a substantial change in surface potential occurs, and the target current is higher following the breakdown. The last three breakdowns, however, display a substantially different character. Very little, if any, change in surface field occurs as the result of these breakdowns. Also, in the last three breakdowns, the target current is affected only slightly. Significantly, the erosion of mirror panels of the sort shown in Fig. 2 occurred only at current levels sufficiently high that breakdowns typified by the last three pulses in Fig. 5 occurred. At lower current densities, the damage was confined largely to the erosion of RTV from the joints between adjacent mirrors. The existence of such a threshold for breakdown characteristics and mirror damage indicated that an accurate knowledge of the peak magnitudes of spacecraft-charging currents is very important. This result also indicates that in laboratory simulations it is important to include provisions for applying peak currents higher than the mean current densities reported.

To gain insight into the possible electrical signals generated in vehicle circuitry as the result of thermal control/mirror voltage breakdowns, the setup shown in Fig. 6 was assembled. A bare wire 12 in. long was placed parallel to the edge of the thermal control panel and returned to the

STUDIES OF VOLTAGE BREAKDOWN PROCESSES

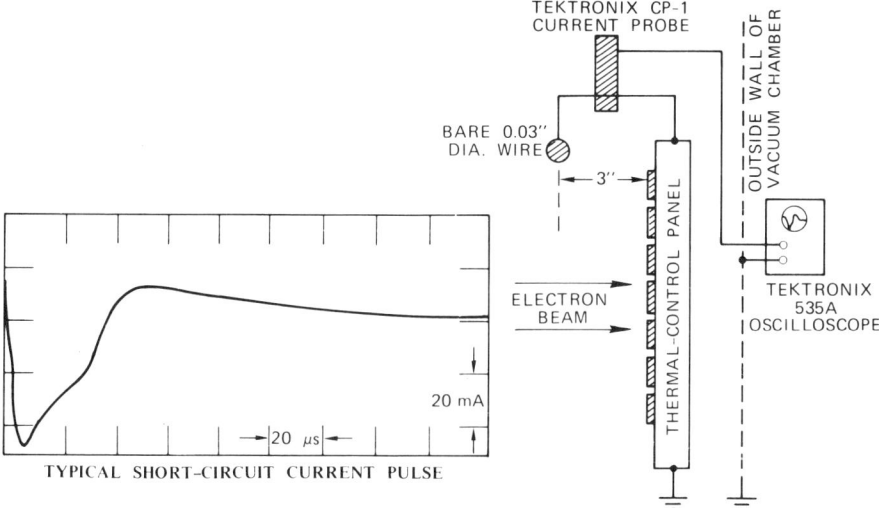

Fig. 6 Short-circuit current probe measurement setup.

panel at one end. A current transformer in the ground lead permitted the measurement of the short-circuit current induced in the resulting electric dipole antenna. It was found that the mirror breakdowns generate substantial currents in the antenna. The peak amplitudes of these current pulses were approximately 60 mA and their durations were of the order of 30 µsec. If the pickup wire of this experiment had been a part of a circuit terminated with a 50-Ω impedance to ground, a 3-V-peak signal could have been induced in the system by the breakdown. For higher impedance circuits, the pulse voltage could be proportionally higher. It should be observed also that the breakdown in Fig. 6 generated a negative-going pulse in the antenna. Physically, this implies that negative charge was driven toward the antenna as the result of the breakdown.

When viewing the mirror panel edge-on during a discharge, it appeared that the breakdown did not consist of a process in which charge was transferred from the mirror surface to the metal substrate. Visually, it appeared that, when a discharge occurred, hot material was ejected to a distance of 1 in. or more above the mirror surface. If the ejected material were charged, the moving charged material would constitute a dipole moment of considerable magnitude and would be very effective in generating high-level signals in nearby receiving systems.

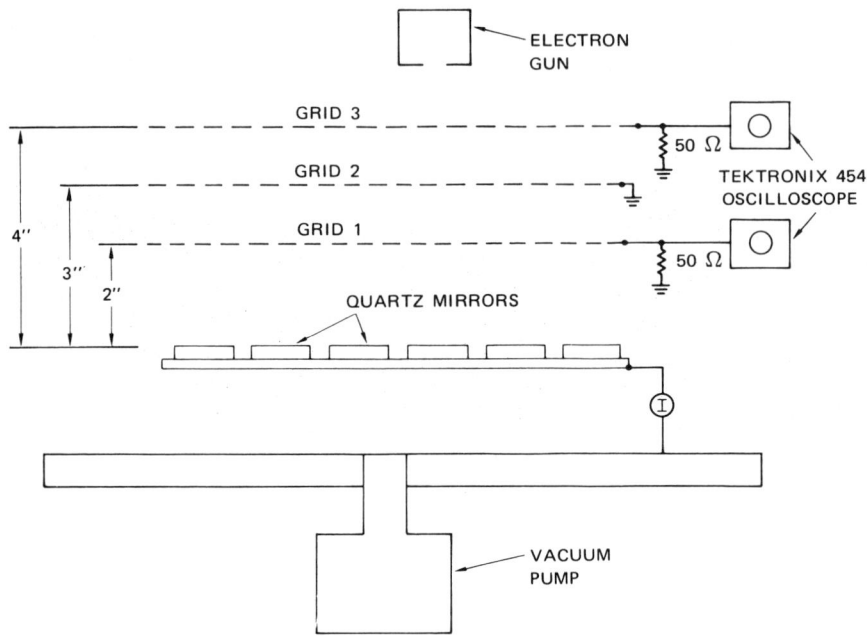

Fig. 7 Laboratory setup for the study of mirror breakdown characteristics.

To investigate the nature of the breakdowns further, the laboratory setup was modified by placing a set of three coarse grids, oriented parallel to the mirror panel, over the panel surface as shown in Fig. 7. The successive distances of the grids from the panel surface were 2, 3, and 4 in. Grid 2 was grounded, and grids 1 and 3 each were connected to the input of an oscilloscope terminated with a 50-Ω resistor. The rationale for this sensor arrangement was that it creates two interaction regions in the space above the mirrors. The first region lies between the grounded mirror panel and the grounded grid, and the second comprises the space above the grounded grid. Charge accumulating in the first region induces a signal in grid 1, whereas charge accumulating in the second region induces a signal in grid 3.

Thus, a signal induced in grid 3 indicates that charged discharge products were expelled to a distance beyond grid 2 (3 in. above the mirror surface). A comparison of the signals induced in grids 1 and 3 determines the degree to which the

charged effluent conditions changed as the material moved
outward from the mirror surface. In general, it was found
that pulses of up to 20-V magnitude were generated in grid 1.
Some of these were associated with breakdowns that visually
appeared to occur almost simultaneously at two places on the
mirror panel. These generally show a break in the slope of
the pulse corresponding to the second breakdown, as shown in
Fig. 8. When a pulse occurred in grid 1, a pulse generally
also occurred in grid 3, indicating that the discharge products
were driven to a distance of over 3 in. from the mirror
surface.

The signals induced in grid 3 have the same gross features
as those induced in grid 1 (i.e., a signal generated by two
breakdowns also consisted of two pulses), but their detailed
structure was sufficiently different from the signals in grid
1 that they could not be ascribed to cross-talk (Fig. 8).
The amplitudes of the signals induced in grid 3 were roughly
half the amplitude of the grid 1 signals, which means that a
considerable fraction of the charged effluent moves beyond
grid 2.

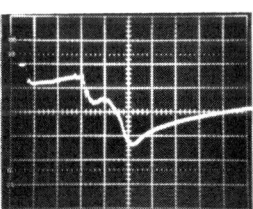

GRID 1 SIGNAL
VERT: 5 V/div
HORIZ: 0.2 µs/div

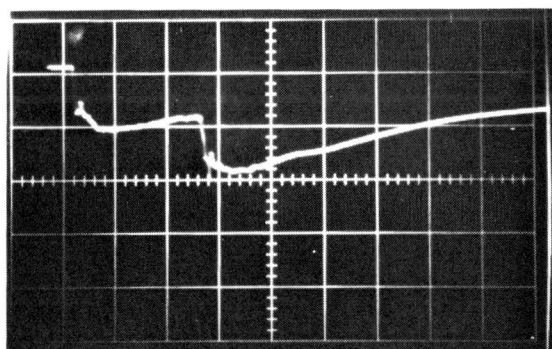

GRID 3 SIGNAL
VERT: 2 V/div
HORIZ: 0.2 µs/div

NOTE: Visual observation indicated two separate flashes

Fig. 8 Investigation of propagation of effluent from mirror
breakdowns.

The polarity of the signals is also of interest. Virtually without exception, the pulses induced by the breakdowns were negative-going. This means that negatively charged material generated in the breakdowns was repelled from the mirror surface by the negative charge deposited by the electron beam. This result is in good agreement with the pulse polarity observed in Fig. 6.

Time and equipment available for this study did not permit the velocity of the charged particles to be determined. Also, it was not possible to perform experiments to determine their nature (ions or electrons). Such experiments would be desirable in further studies of the mirror breakdown phenomenon, since such information is required in calculating the noise signals that will be generated by such discharges.

Conclusions and Recommendations

The work reported here demonstrates that quartz thermal-control mirrors under electron bombardment in a vacuum can acquire sufficient surface charge to cause voltage breakdown. The breakdown repetition rate is proportional to the current density of the illuminating source. Inspection of the thermal-control panels after a period of breakdowns indicates that physical damage to the RTV sealant and even to the mirrors has occurred.

The breakdowns can induce substantial signals in nearby electrical wiring. A brief investigation into the physical nature of the breakdowns indicates that, in the course of a discharge, negative charge is expelled to a distance of over 3 in. from the mirror surface. This result means that the effective dipole length of the discharge is not the thickness of a mirror panel, but is several inches. Thus, such breakdowns should be expected to be energetic sources of noise pulses of considerable time duration.

Additional work could be done profitably to clarify the physical processes occurring during the breakdowns. This information is needed to permit accurate estimates of the noise signals produced by the breakdowns. In situ measurements would be desirable on a synchronous orbit satellite to verify the laboratory results and to look for unanticipated processes.

References

[1] Nanevicz, J. E., and Hilbers, G. R., "Titan Vehicle Electrostatic Environment," TR AFAL-TR-73-170, Contract F33615-70-C-1406, Project 8428, July 1973, Stanford Research Institute, Menlo Park, Calif.

[2] Nanevicz, J. E., Adamo, R. C., and Scharfman, W. E., "Satellite-Lifetime Monitoring," Final Report, Contract F04701-71-C-0130, P. O. 126192, Project 2611, March 1974, Stanford Research Institute, Menlo Park, Calif.

[3] Shaw, R. R., Nanevicz, J. E., and Adamo, R. C., "Observations of Electrical Discharges Caused by Differential Satellite Charging," -- published elsewhere in this volume.

[4] DeForest, S. E., "Spacecraft Charging at Synchronous Orbit," Journal of Geophysical Research, Vol. 77, Feb. 1972, pp. 651-659.

[5] DeForest, S. E., "Electrostatic Potentials Developed by ATS-5," 1972 Lightning and Static Electricity Conference Papers, Air Force Systems Command, Wright-Patterson Air Force Base, Ohio, AFAL-TR-72-325, 1972, pp. 15-165.

EVIDENCE FOR A NEW DISCHARGE MECHANISM FOR
DIELECTRICS IN A PLASMA

A. Meulenberg Jr.*

COMSAT Laboratories, Clarksburg, Md.

Abstract

This paper proposes a bilayer discharge immediately within the outer surface layer of a dielectric in a plasma, rather than a discharge through or between spacecraft surface components. The bilayer consists of electrons that penetrate into the bulk and positive charge left on the surface as a result of electron depletion by secondary emission. If the bulk resistivity is sufficiently high and the incident electron flux from the plasma is sufficiently intense, then the potential gradients will exceed the breakdown field of the material. Simple modeling of the system shows good agreement with the laboratory results, suggesting that this form of discharge, in fact, does occur at magnetic substorm plasma densities and energies.

Introduction

Correlations between magnetic substorms, spacecraft-charging, electrical discharges, and synchronous satellite

---

Presented as paper SA70 at the American Geophysical Union Spring Annual Meeting, June 16—19, 1976, Washington, D.C. The author would like to thank E. S. Rittner for encouragement and support, L. H. Westerlund for helpful discussions, and A. G. Revesz for providing background material on dielectrics. This paper is based upon work performed at COMSAT Laboratories under the sponsorship of the International Telecommunications Satellite Organization (INTELSAT). Views expressed in this paper are not necessarily those of INTELSAT.
 *Member of the Technical Staff, COMSAT Laboratories.

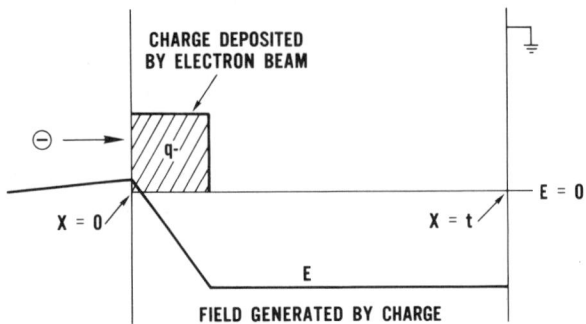

Fig. 1 Assumed deposited-charge profile and resulting fields for low-energy electrons incident from the left.

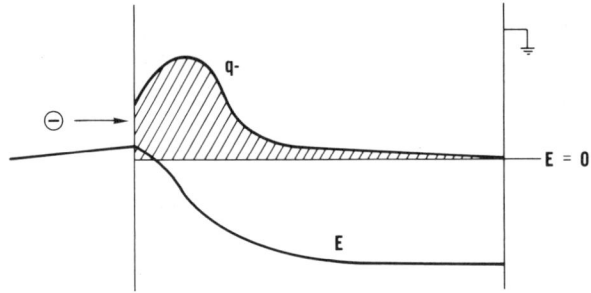

Fig. 2 Charge density and resulting fields from low-energy electrons incident on the left, assuming finite bulk resistivity and small secondary-electron emission.

anomalies have been well established.[1-4] Conventional theory[5] requires sunlight, in addition to the magnetic substorms, to create the potential gradients necessary to generate electrical discharges. However, many discharges and operational malfunctions have been observed on spacecraft without substorms or in "quiet" sections of the synchronous orbit, as well as during eclipse.[6,7] An explanation that goes beyond the conventional theory is required. This paper will seek to provide the foundation for an extended discharge theory.

Conventional Model

Since other papers have covered the electrical breakdown between satellite components (e.g., Ref. 5), this paper will discuss the breakdown within spacecraft dielectrics. This is a complicated problem with many inadequately known variables. As a starting point, a simple model of monoenergetic electrons, normally incident on a dielectric with a step-function charge deposition, will be used. Figure 1 shows the charge and field distribution for this model with electrons coming in from the left and a grounded conductor on the back. Additional conditions assumed in this model are no secondary emission and no charge redistribution in the dielectric (i.e., infinite resistivity). The system will act as a capacitor with the back surface of the dielectric grounded (a source or sink for electrons). The electric field from the electrons deposited in the front surface will force electrons to drain from the back surface, leaving a positive charge that reduces the field E outside the dielectric.

Electrons within the front layer then are allowed to redistribute under the influence of the field within the dielectric. The negative field in Fig. 1 indicates that electrons will migrate to the right in the dielectric. After equilibrium is established between the incoming beam and the drainage currents, the charge distributions and fields would be as shown in Fig. 2. This figure was constructed with the assumptions of finite bulk resistivity, some secondary emission from the surface, and fringing fields resulting from the finite dimensions of the scattering chamber. It should be noted that, in the conventional model (Fig. 2), the strongest field is near the back surface. Breakdown, if it occurs, would be expected to start at this point and consist of a discharge of electrons to the back surface.

Present Work

Experimental

A laboratory experiment was set up using 8-keV electrons normally incident on various dielectrics common to spacecraft (Kapton, Mylar, and Teflon). Thin sheets of these materials were placed on a conductive surface which was connected to a Keithley electrometer set on "fast" response to insure a low-impedance path to ground. Electrons were generated by a hot filament, focused, and bent through a magnet to form a uniform beam without contamination of light or negative ions. This beam was swept onto the target, and quasi-equilibrium was established within 1-4 sec.

Figure 3 is representative of the current collected from the conductive surface. Mylar and Teflon exhibited the "ringing" shown in Fig. 3 and a continued slow reduction in the collected current, but Kapton did not. After a length of time that depended on the beam current, material thickness, and past history, a pulse was observed. From the previous model, a pulse in the same direction as the incident beam current was

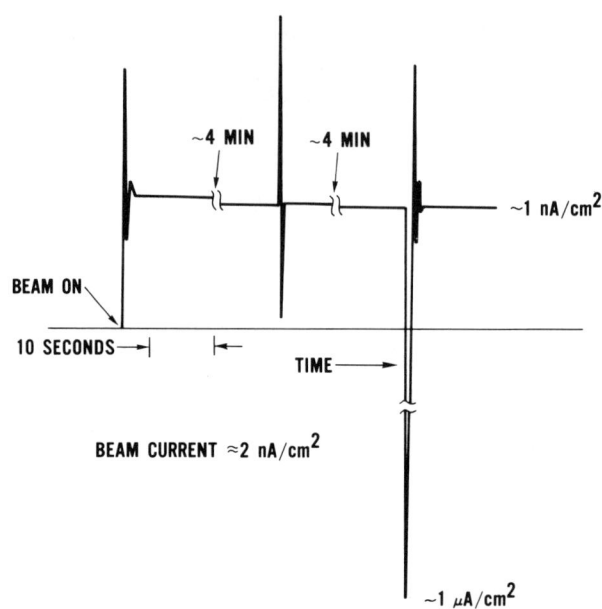

Fig. 3 Electron current collected from conductive back surface of a dielectric with low-energy electrons incident on the front.

expected. These pulses were in fact observed, but larger pulses in the opposite direction also were observed. These reverse pulses occurred in all samples that had not been heavily irradiated previously.

At a beam current of 1 na/cm^2, these pulses appeared at a rate of 4-8/hr. At increased beam currents, the pulse rate was higher until the material began to "fatigue," at which point the pulsing slowed and eventually stopped entirely. This effect is thought to be a result of decreased resistivity resulting from damage to the material. The possibility of surface effects cannot be discounted, since long exposure to an ion pumped vacuum ($<10^{-6}$ Torr) prior to irradiation also reduced the pulse rate for a given beam current. After a significant effort to show that these reverse pulses were not artifacts of the experiment setup or instrumentation, the model was reexamined.

Theory

The reduced electron density at the front surface in Fig. 2 is a result of secondary-electron emission. These secondary electrons are removed from a very shallow layer relative to the electron penetration depth. If electrons are removed from a layer one-tenth the thickness of the layer in which the beam is deposited, then a secondary-emission coefficient of $\delta = 0.1$ would imply that there is no net gain of electrons in this front region. As the secondary-emission coefficient increases, the electron concentration near the surface decreases further, and depletion actually can result. This positively charged depletion region will draw electrons, deposited deeper in the material, toward the surface. The extent and magnitude of electron depletion at equilibrium depend on the nature (energy and angular distribution) of the incident radiation; the characteristics (resistivity, charge mobility, and energy of ionization) of the dielectric; and the secondary-emission coefficient under the measurement conditions.

If the secondary-emission coefficient δ of the front surface is >0.5, it becomes apparent that more current will flow from the negatively charged layer to the front surface than to the rear. If the bulk resistivity is high, then high fields are required to establish this flow. (Sellen[8] independently derived these high fields for $\delta = 1$ in an excellent description of the interaction of materials with the low-energy space radiation environment.) With a grounded back contact, the only way to create high fields in the neighborhood of the front surface is to have a sizable positive charge on that surface. This positive surface charge (due to electron depletion) arises because the electrons, equivalent to nearly one-half of the beam current, are removed from a very shallow layer, including the surface, and most of the electrons in the beam are deposited in the bulk region for a distance up to the electron range.

Figure 4 is an enlargement of the front region of the dielectric in Fig. 2 and represents the equilibrium condition for a more realistic value of the secondary-emission coefficient (0.5). Note that the positive charge on the front surface, as a result of secondary emission, has shifted the field curve upward, causing a large field near the surface and reducing the field inside the sample. Electrons now are as likely to move toward the front as to the rear. Breakdown of the dielectric thus will occur near the front surface, with electrons going to and through this surface. (Even if electrons do not leave the surface, a sharp pulse will be induced in adjacent conductors.[9]) The field profile and consequent charge redistribution is therefore the key to the origin of reverse pulses resulting from discharges of the formed bilayer and probably to the understanding of secondary-emission characteristics in this system.

The actual charge density profile in the dielectric depends on beam energy and intensity, material resistivity,[10]

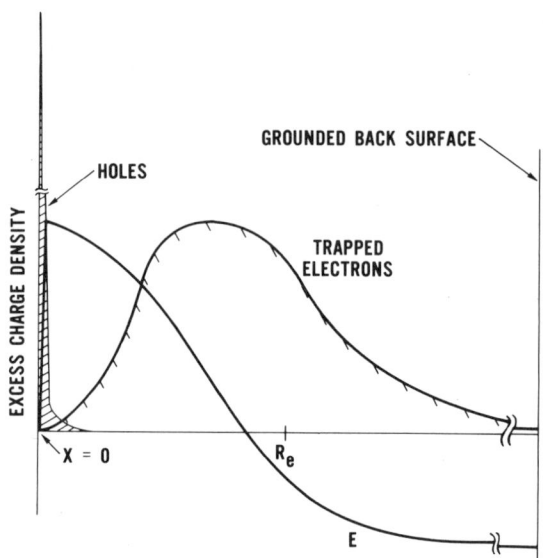

Fig. 4 Detail of Fig. 2 assuming a secondary-emission coefficient of $\delta \geq 0.5$.

"hot" electron conductivity, secondary-emission characteristics, and back-surface conductor potential. The energy stored in the bilayer depends on the total charge and its density profile, which in turn determine the field intensity and shape. The power dissipated in a discharge depends on the energy density of the bilayer, the rate at which the breakdown is propagated, and the area of surface involved. Since the bilayer is of the order of 1-μm-thick, the capacitance, energy density, and local discharge rates are very high. Calculations show that the capacitance per unit area of this bilayer is in the nF/cm^2 range and is a function of the charge profile. The bilayer energy density, for breakdown fields of ~100 V/μm, can reach the order of 10^{-4} J/cm^2.

A discharge in the surface layer can boil off constituent atoms and molecules of the dielectric. (Experimental evidence of this aspect has been shown by Nanevicz.[11]) The resulting plasma may act as a dipole antenna, thereby increasing the output and directionality of the rf emission from the discharge. In addition, this material could deposit on other spacecraft surface components, hence degrading their characteristics and perhaps forming new arcing sites.[12]

There is already a body of data on the polarization of dielectrics by radiation and dielectric discharge in capaci-

A NEW DISCHARGE MECHANISM IN DIELECTRICS 243

tors.[13-19] These data support the internal breakdown model, and in fact, some indicate spontaneous discharge occurrence. They also show that slow discharge of a dielectric is enhanced by thermal and/or photoelectric excitation.

So far, only monoenergetic, normally incident, electron beams have been considered. It is apparent that the charge distribution in the dielectric will be altered significantly if the direction and/or energy distribution of the beam is different. Electrons at non-normal incidence angles induce higher secondary-emission coefficients and will be deposited less deeply. A monoenergetic beam of electrons incident at a shallow angle will produce a steeper field profile than a similar beam at normal incidence, causing breakdown to occur sooner. For an energy spectrum of electrons, as opposed to a monoenergetic beam, the field profile will be steeper, secondary emission will normally be greater, and breakdown will occur again sooner. If breakdown occurs sooner, then the energy density of the bilayer will be lower. This smaller discharge may not spread as fast or as far as a large one; hence, many "micropulses" may occur rather than a single massive discharge.

Comments

Work, both experimental and theoretical, must be continued to determine the material parameters necessary to quantify the details of this mechanism, but experiments and calculations performed so far have supported it. Some brief observations can be made from these various experiments:

1) The large reverse pulses and intense rf pulses observed indicate that a fast discharge of a large area occurs, and that electrons from the bulk do not stop at the surface but continue on out of the dielectric. It is quite probable that surface atoms are carried along to form a plasma, which acts as an "antenna," thereby increasing the rf coupling to spacecraft elements.

2) This mechanism will occur during eclipse, since the breakdown fields are generated within the dielectric by the electron portion of a hot plasma alone. The lack of sunlight, which pins the spacecraft frame at low potentials, actually will increase the fields near the surface. Both the higher fields and the decreased temperature during this period increase the discharge probability for a given electron environment.

3) Sunlight will decrease the probability of this type of discharge by thermally and photoconductively decreasing the

bulk resistivity of the dielectric. In spacecraft with conductive layers attached to the spacecraft frame, sunlight will lower the dielectric front-surface fields (and raise the back fields) by "grounding" the spacecraft. On the other hand, the ultraviolet light will help to remove surface electrons by photoemission[17] and also may act as a discharge "trigger."

4) The energy spectrum of isotropic electrons in a space plasma will increase the field gradients near the surface, relative to those created by a unidirectional beam. Localized discharges then could appear as surface "scintillation", rather than as major arcs resulting from a deeper, more uniform field.

5) Thin metallic surface layers should prevent the "scintillation" problem, but will not prevent major arcs created by undirectional or higher-energy isotropic electron beams.

Summary

This paper has proposed a bilayer model for dielectric discharge in a plasma which provides for intense discharges that do not depend on the spacecraft being in sunlight, or for that matter, even in the vicinity of major magnetic substorms. This model covers many discharge-associated anomalies not previously explained by conventional theory. However, better values for the various material parameters are necessary to determine the probability of bilayer discharge in a dielectric, relative to that of discharge from the front surface to the back of a dielectric, or between adjacent conductive layers in the spacecraft outer components.

References

[1] Rosen, A., "Spacecraft Charging: Environment Induced Anomalies," AIAA Paper 75-91, Jan. 1975, Pasadena, Calif.

[2] McPherson, D. A., Cauffman, D. P., and Schober, W., "Spacecraft Charging at High Altitudes—The SCATHA Satellite Program," published elsewhere in this volume.

[3] Deforest, S. E., "Electrostatic Potentials Developed by ATS-5," Photon and Particle Interaction with Surfaces in Space, edited by R. J. L. Grard, D. Reidel Publishing Co., Dordrecht, Holland, 1973.

[4] Deforest, S. E., "Spacecraft Charging at Synchronous Orbit," Journal of Geophysical Research, Vol. 77, Feb. 1, 1972, pp. 651-659.

[5] Inouye, G. T., "Spacecraft Charging Model," *Journal of Spacecraft and Rockets*, Vol. 12, Oct. 1975, pp. 613–620.

[6] Nanevicz, J. E., Adamo, R. C., and Scharfman, W. E., "Satellite Lifetime Monitoring," Stanford Research Institute, Menlo Park, Calif., Project 2611, 1974.

[7] Robbins, A., Royal Aircraft Establishment, private communication, 1975.

[8] Sellen, J. M., Jr., "Electrical Equilibrium of Conducting and Insulating Materials in the Presence of Energetic Electrons and Ultraviolet Light," TRW Systems, Inc., Redondo Beach, Calif., 4351.3.74-44, 1974.

[9] Lindmayer, J., "Current Transients in Insulators," *Journal of Applied Physics*, Vol. 36, Jan. 1965, pp. 196–201.

[10] Hoffmaster, D. K. and Sellen, J. M., Jr., "Electron Swarm Tunnel Measurements of Kapton Bulk Resistivity at High Electric Stress Levels," TRW Systems Inc., Redondo Beach, Calif., 4351.3.74-39, 1974.

[11] Adamo, R. C. and Nanevicz, J. E., "Spacecraft Charging—Studies of Voltage Breakdown Processes on Spacecraft Thermal Control Mirrors," published elsewhere in this volume.

[12] Nanevicz, J. E., "Malter Discharges as a Possible Mechanism Responsible for Noise Pulses Observed on Synchronous-Orbit Satellites," published elsewhere in this volume.

[13] Murphy, P. V. and Ribeiro, S. C., "Polarization of Dielectrics by Nuclear Radiation, Part I—Release of Space-Charge in Electron Irradiated Dielectrics," *Journal of Applied Physics*, Vol. 34, July 1963, pp. 2061–2063.

[14] Murphy, P. V. and Gross, B., "Polarization of Dielectrics by Nuclear Radiation, Part II: Gamma-Ray-Induced Polarization," *Journal of Applied Physics*, Vol. 35, Jan. 1964, pp. 171–174.

[15] Monteith, L. K., "Trapping and Thermal Release of Irradiation Electrons from Polyethylene Terephthalate Films," *Journal of Applied Physics*, Vol. 37, June 1966, pp. 2633–2639.

[16] Monteith, L. K. and Hauser, J. R., "Space-Charge Effects in Insulators Resulting from Electron Irradiation," *Journal of Applied Physics*, Vol. 38, Dec. 1967, pp. 5355–5365.

[17]Powell, R. J. and Derbenwick, G. F., "Vacuum Ultraviolet Radiation Effects in SiO$_2$," IEEE Transactions on Nuclear Science, Vol. NS-18, Dec. 1971, pp. 99—105.

[18]Ausman, G. A., Jr. and McLean, F. B., "Electron-Hole Pair Creation Energy in SiO$_2$," Applied Physics Letters, Vol. 26, Feb. 1974, pp. 173—178.

[19]Curtis, O. L., Jr., Srour, J. B., and Chin, K. Y., "Hole and Electron Transport in SiO$_2$ Films," Journal of Applied Physics, Vol. 45, Oct. 1974, pp. 4506—4513.

MALTER DISCHARGES AS A POSSIBLE
MECHANISM RESPONSIBLE FOR NOISE PULSES
OBSERVED ON SYNCHRONOUS-ORBIT SATELLITES

J. E. Nanevicz[*] and R. C. Adamo[†]
Stanford Research Institute, Menlo Park, Calif.

Abstract

Differential charging of insulated surfaces on synchronous-orbit satellites has been postulated as a mechanism capable of generating voltage breakdowns of these surfaces, which in turn lead to rf noise pulses and to the generation of gaseous products capable of causing degradation of optical surfaces on the vehicle. Although surface potentials of several hundred volts in sunlight and several kilovolts in the dark are possible during a geomagnetic storm, experiments on a synchronous-orbit satellite indicate that breakdown pulses occurred when surface potentials in the vicinity of the surface-potential sensor were of the order of 50 V. A likely mechanism to account for these breakdowns is a process described by Malter, which involves current flowing to a metal surface on which a thin insulating film has been formed. Current flowing through the film will generate a potential gradient across the film high enough to produce semiconductor-junction breakdown. Discharges of this type have been produced in the laboratory on Iridite-15-coated magnesium sections of sample thermal-control panels at potentials as low as 30 V. Substantial rf noise pulses were

---

Presented as Paper SA71 at the American Geophysical Union Spring Annual Meeting, Washington, D.C., June 16-19, 1975.
[*]Program Manager.
[†]Research Engineer.

found to accompany these breakdowns, the insulating surface was destroyed, and some of the metal substrate was vaporized.

Introduction

In connection with the operation of synchronous-orbit satellites, it was found that anomalous behavior of systems occasionally occurred, and that this behavior was consistent with the occurrence of discharges or current surges on the satellite. In addition, it was found that a statistical correlation existed between anomalous temperature rise on certain satellites and geomagnetic indices. To investigate any possible connection between thermal behavior and satellite surface-charging, an instrumentation system developed for the study of rocket charging[1] was repackaged and installed on a non-NASA synchronous-orbit satellite.[2] When sensitivity levels were being established for the satellite instrumentation, results of orbital measurements[3,4] and calculations based on laboratory experiments[5,6] were used in a series of laboratory experiments designed to simulate the charging of the vehicle's thermal-control surfaces.[2,7] From this work, it was concluded that breakdown pulses should be observed when the potential difference between the thermal-control surface and the satellite structure reached 2000-3000 V. DeForest's measurements[3,4] indicated that such potential differences indeed occurred during energetic particle injection events associated with magnetic substorms when the satellite was eclipsed by the Earth.

When satellite data became available, it was found that rf noise pulses occurred throughout the orbit,[2,8] and that the thermal-control surface potential was 50 V or less when they were observed.[2] This result indicated that although electrical breakdowns associated with differential-charging potentials of thousands of volts were possible, a search should be made for other mechanisms capable of generating electrical breakdowns when the differential potential is of the order of tens of volts. Meulenberg[9] observed that the electron penetration, subsurface charge storage, and resulting breakdowns observed by experimenters irradiating dielectric materials with high-energy ($\approx$ 1 MeV) electron beams[10,11] also can occur when satellite dielectric surfaces are irradiated with electrons with energies of tens of kilovolts.

The authors' experience with dielectric in plasmas indicated that breakdowns can occur when charge extracted from a plasma is deposited on thin films of oxide and other materials deposited on metal electrodes within the plasma region. In particular, these breakdowns appear as scintillations of light on one or the other of the electrodes. Since the oxides or other surface films on the electrode are often insulating or only partially conducting, charge from the ionized gas can accumulate on the film surface and develop very high dc field strengths in the film. When the dielectric strength of the film is exceeded, the film breaks down, and a visible flash or scintillation is observed. These scintillations, which are believed to be the same as those described by others in connection with the Malter effect,[12-16] have been observed on the negative electrode when brass or aluminum electrodes were used. With copper electrodes, however, the scintillations are observed on the positive electrode. It is postulated that the peculiar behavior of copper in this respect results from the fact that the copper cuprous oxide junction offers little resistance to the flow of electrons from the copper to the oxide, but it impedes the flow of electrons from the oxide to the copper. Electrons or negative ions thus can accumulate on the oxide film and distort the field at the electrode surface of copper electrodes, in much the same manner as positive ions are purported to accumulate on the oxides of aluminum and other materials displaying the Malter effect.

Scintillations have been observed by the authors with bias potentials as low as 20 V. Preliminary experiments in which polystyrene films were deposited on carefully cleaned stainless steel substrates indicated that scintillations could be induced from films as thin as 500 Å. It should be observed that, with this film thickness, electron penetration of the entire film is possible with energies of only 1.2 kV.[17] Thus, scintillation breakdown is truly a low-energy phenomenon.

Experiments and Results

To lend relevance to the experiments, it was decided that a material commonly employed in satellite construction should be used as the test sample. Accordingly, several magnesium panels were treated with Iridite 15, which is used commonly as the passivating material in satellite construction.

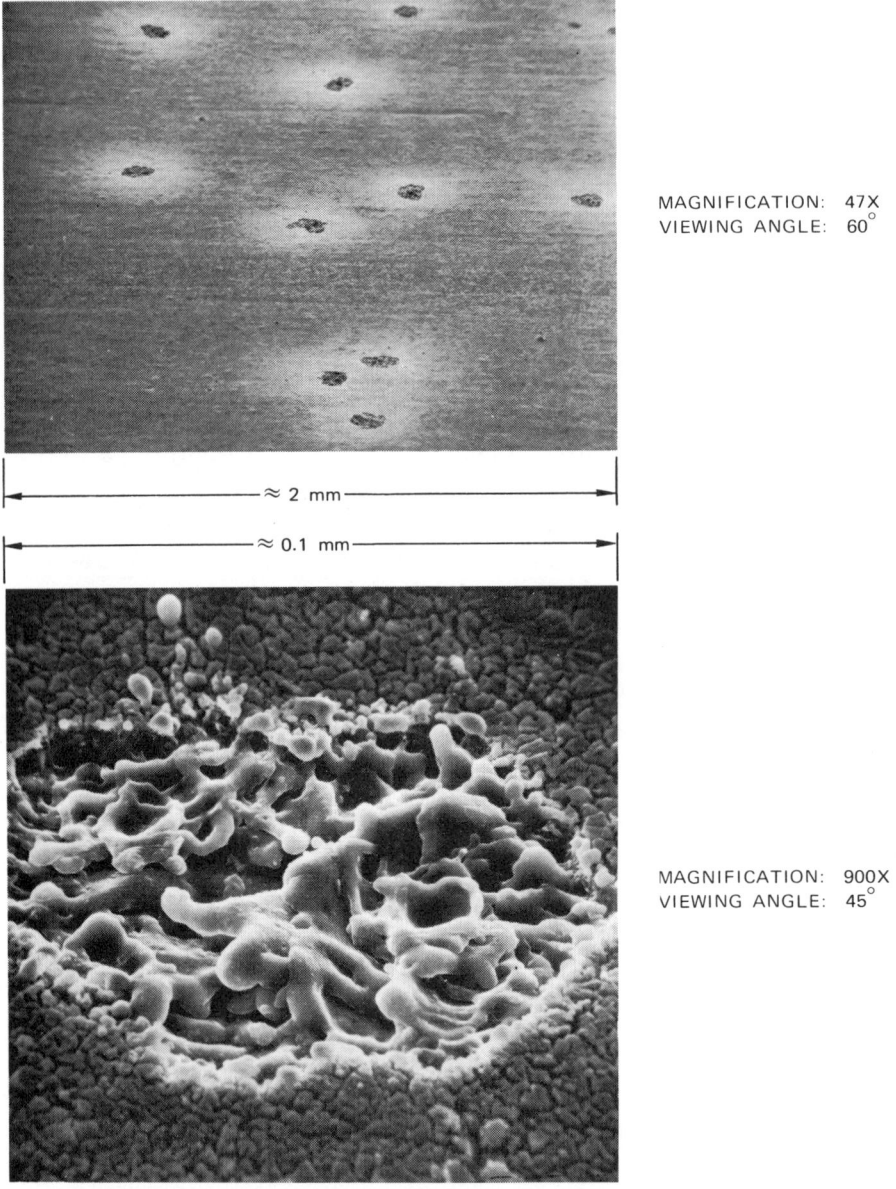

Fig. 1 Physical effects of low-voltage discharges produced during laboratory tests.

In the first set of experiments, the sample was installed in a vacuum chamber equipped with an rf-discharge plasma source. The sample was biased 50 V negative with respect to the metal structure of the vacuum chamber, and scintillation discharges were observed to occur.

The results of a scanning electron microscope examination of the surface following the scintillation discharges are shown in Fig. 1. It can be seen from this figure that a considerable amount of surface material, which may cause contamination of nearby thermal control surfaces, has been liberated by these discharges. Similar results were obtained in experiments with polystyrene films deposited on a stainless steel panel.

In the next set of experiments, some of the mechanisms responsible for the generation of the radio noise observed in connection with scintillation discharges were studied. In working with scintillation discharges, it was observed that when viewing the test panel edge-on, the luminous portion of the breakdown appeared to extend to a considerable distance above the panel surface, indicating that discharge products were blown out from the surface by the discharge. (This is certainly suggested in Fig. 1, although there is no indication of the distance to which the material is carried.) If it were possible to show that the ejected material is moved to a distance of several inches from the surface, and if the material were found to be electrically charged, then one could conclude that a mechanism existed for generating high-amplitude long-duration noise pulses.

The setup used for the studies of scintillation discharge electrical characteristics is shown in Fig. 2. A treated magnesium panel was placed in a vacuum chamber and biased 50 V negative with respect to the metal structure within the chamber. Plasma was generated by means of an rf discharge contained within a grounded screen container. A Mylar sheet with a 1-in.-diam hole was placed over the magnesium test panel to restrict the region where scintillation discharges could occur to the region of the hole. A pair of induction rings with three associated guard rings were positioned in line over the center of the hole. With this arrangement,

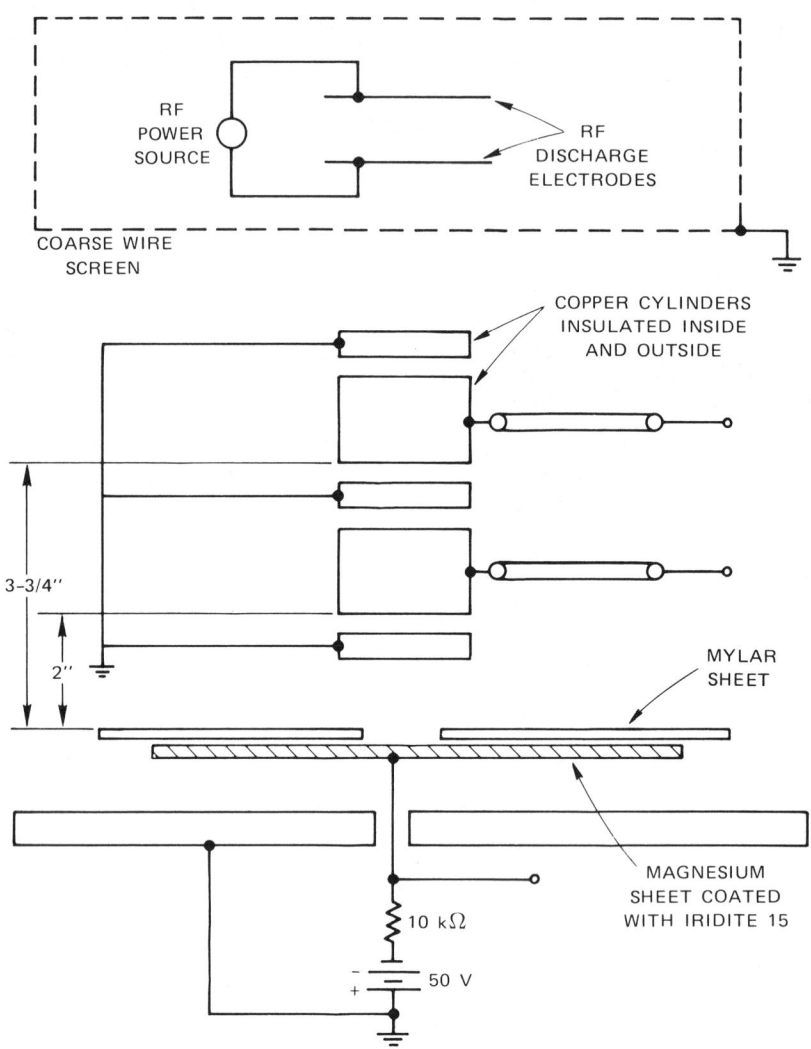

Fig. 2 Laboratory setup for the study of Malter breakdown characteristics.

charged particles, expelled upward, pass through the cylindrical induction rings and induce on the ring a charge equal to the net charge in the region of material enclosed by the cylinder. The input resistance of the oscilloscope channel connected to each of the induction rings was 1 MΩ, while the combination of cable capacitance and amplifier input capacitance was 200 pF. Thus, the system had an input RC time of 200 μsec. This means that for the first 10 μsec or so, very little error results from calculating the enclosed charge q from

$$q = CV \tag{1}$$

where C = total input capacitance = 200 pF and V = observed oscilloscope input voltage. This arrangement allows one to make estimates regarding the extent of the column of ejecta, the velocity of the ejected particles, and the charge density within the column. (Although not shown in Fig. 2, the entire electrode structure was covered with a grounded shield so that charges passing the outside of the structure would not induce signals in the electrodes.) Results of typical experiments are shown in Fig. 3. Here it is evident that negatively charged particles were generated by the discharge, and that these charged products were driven to at least the position of the upper induction ring (3 3/4 in above the surface of the test panel). (The two oscillograms in this and the succeeding two figures are examples of the different discharge types observed under identical conditions.) In the upper oscillogram, the upper trace gradually returned to zero following the negative pulse, whereas in the lower

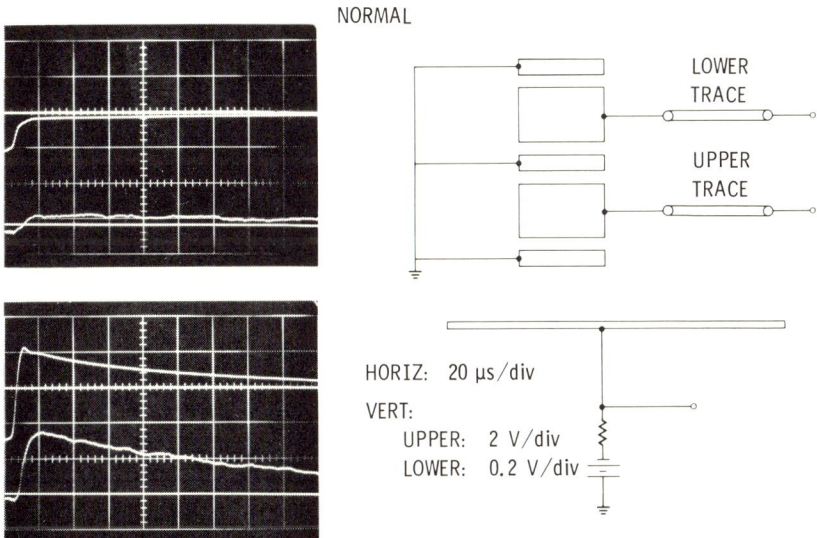

Fig. 3 Experiments demonstrating distance to which charged products are expelled.

oscillogram, a positive overshoot occurred. It is not clear at this time what significance should be attached to the behavior following the initial negative pulse. The occurrence and magnitude of overshoot varied from breakdown to breakdown, and the behavior of the signals from the two electrodes in this time period was not well correlated. Accordingly, since the late-time behavior may be an artifact of the experimental setup, attention will be restricted to the data contained in the initial part of the pulse.

Returning to Fig. 3, it is evident that no discernible time delay occurs between the arrival of the charged products at the lower and upper induction rings. Oscillograms taken at higher sweep speeds indicate that the time difference was less than 0.1 μsec. (Time did not permit a more accurate determination. It is hoped that this will be done in the near future.) Since spacing between the two electrodes is roughly 5 cm, this means that the velocity $v > 5 \times 10^4$ m/sec. Using the relationship

$$v = \sqrt{2eV/m} \qquad (2)$$

where v = particle velocity, e = electronic charge = 1.6×10^{-19} C, V = applied potential, and m = particle mass, we find that to accelerate an electron to this velocity requires only 7×10^{-3} V. Thus the particles responsible for the negative-going pulse could be electrons.

Conversely, if we assume that the full 50 V applied potential is available to accelerate particles away from the source of the breakdown and again, $v > 5 \times 10^4$ m/sec, we can rewrite Eq. (2) and find the maximum mass of the negatively charged particle. When this is carried out, we find that the particle must have an atomic weight less than 3.8. Thus we must conclude that the particles responsible for the initial negative-going pulse are electrons.

From the top trace of the upper oscillogram in Fig. 3, we find that the amplitude of the negative pulse is V = 2.2 V. Substituting this value in Eq. (1), we find that the charge contained in the cylindrical electrode is q = (200×10^{-12}) $(2.2) = 440 \times 10^{-12}$ C. Since the cylinder is 1 in. long, this means that the charge per unit length 2 in. above the test panel is $\sigma = q/\ell = 1.7 \times 10^{-8}$ C/m (for upper oscillogram).

Next, it is interesting to compare the amplitudes of the signals generated in the upper and lower induction rings. From the upper oscillogram of Fig. 3, we find that the ratio of amplitudes is 2.2/0.04 = 55. If the particles followed straight-line trajectories from a point source, we would expect the current density to vary inversely as the square of the distance from the panel, so that the ratio of amplitudes would be 4. This great disparity indicates that the particles are being dispersed more rapidly, possibly as the result of coulomb forces within the cloud of discharge products.

Let us now calculate the magnitude of the field at the periphery of the column of charge as it reaches the lower induction ring. From Gauss's law for a cylinder of unit length, we can write

$$2 \pi r \epsilon_o E = \lambda \qquad (3)$$

where ϵ_o = dielectric constant = $1/36\pi \times 10^9$, E = field intensity, r = radius of cylinder, and λ = charge per unit length along cylinder. If we solve Eq. (3) for E, and note that the induction ring has a diameter of 1 in., we find that, for $\lambda = 1.7 \times 10^{-8}$ C/m (corresponding to the upper oscillogram in Fig. 3), the field intensity at the edge of the column is E = 24 kV/m. This is a substantial field and indeed will cause a rapid divergence of the electrons in the discharge product cloud as was suggested in the preceding paragraph.

The remaining experiments were conducted largely to verify that the instrumentation was functioning as planned, and that the results obtained were not the consequence of some unexpected interactions between the instrumentation system and the physical processes. The first of these, shown in Fig. 4, was designed to verify that the signal induced in the upper induction ring was generated by particles passing upward through the ring. A polystyrene foam cork was placed inside the tube midway between the two induction rings where it would block particles coming up the tube. Since the foam is

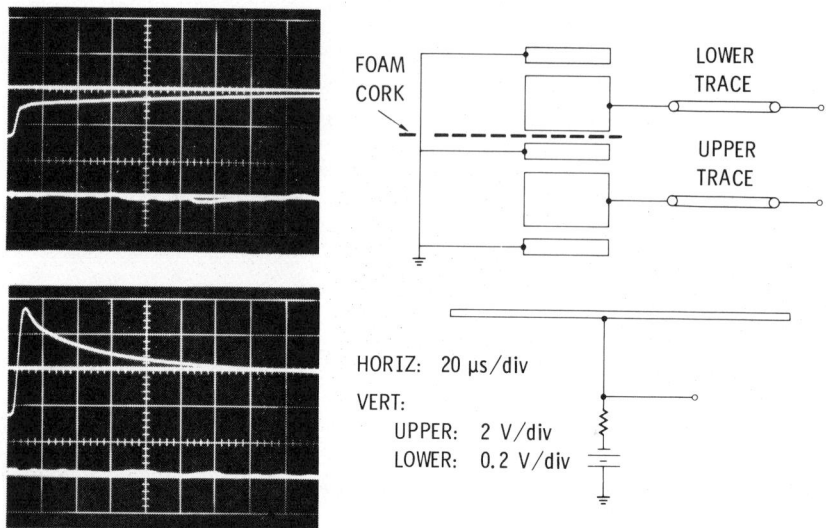

Fig. 4 Setup to demonstrate that signal in upper induction ring is generated by moving particles.

transparent to rf signals, it does not modify any stray rf coupling paths that might be responsible for the signals induced in the upper induction ring. Also, the cork does not affect particles entering the top of the tube or passing by the outside of the tube. It is evident from Fig. 4 that blocking the tube in the middle eliminated the signal in the upper induction ring.

The setup shown in Fig. 5 was intended primarily to provide a scheme for triggering the oscilloscope, even when no signal was induced in the induction rings. To accomplish this, the signal generated in the test sample was applied to the oscilloscope. Since this signal is displayed, it will be interesting to see what it indicates regarding the breakdown process. First, it is interesting to note that a positive-going pulse is generated. This is consistent with electrons leaving the test panel and passing through the induction ring structure.

Next, it will be of interest to use the test panel current data to estimate the charge transported. Because the 10-kΩ resistor in series with the bias supply is effectively the load resistance for the system, the RC time is

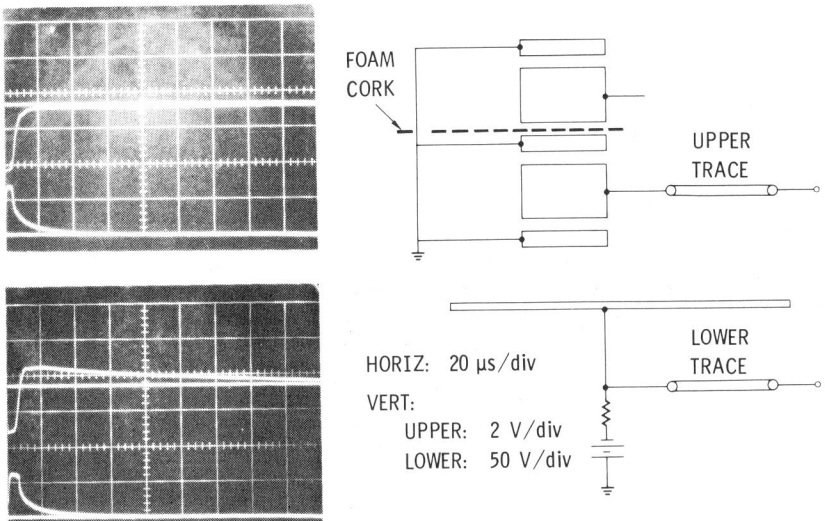

Fig. 5 Provision to trigger oscilloscope from signal generated in test sample.

$10^4 (200 \times 10^{-12}) = 2 \times 10^{-6}$ sec. This means that the voltage induced in the oscilloscope is given by

$$V = IR \qquad (4)$$

instead of Eq. (1) for times greater than roughly 2 μsec. From the upper oscillogram, we find that V = 75 V so that $I = 7.5 \times 10^{-3}$ A. We note that the duration of the pulse $T = 4 \times 10^{-6}$ sec. Substituting these values into

$$Q = IT \qquad (5)$$

we obtain $Q = 30 \times 10^{-9}$ C. It is interesting to note that an individual streamer pulse occurring on a dielectric surface of an aircraft transfers $1 - 1.5 \times 10^{-9}$ C of charge over a distance of several inches.[18,19] Since streamers are known to disable aircraft electronic systems, a process capable of transferring 30 times as much charge over a comparable distance certainly should be capable of generating substantial noise pulses.

Finally, in the setup shown in Fig. 6, it was shown that the signals generated in the lower induction ring were

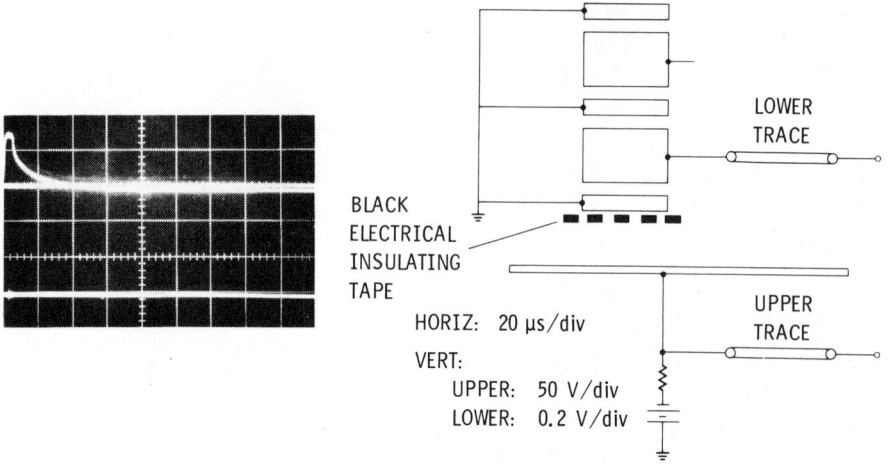

Fig. 6 Setup to demonstrate that signal in lower induction ring is generated by moving particles.

generated strictly by charged particles passing through the ring. In this setup, it was shown that a single layer of black insulation tape placed over the bottom opening of the tube completely eliminated the signal induced in the lower induction ring.

It should be noted that in order to support scintillation discharges on the Iridite 15 - treated magnesium test panels, it was necessary to maintain a current density $J = 600$ nA/cm^2. This is a substantial current density. It should be recalled, however, that the experiments reported here were in the nature of a preliminary exploration, so that only one material was used, and the tests were carried out at room temperature. Recent photoconductivity experiments performed at Stanford Research Institute indicate that the conductivity of a material varies markedly with temperature and illumination.

Thus, by a suitable selection of material at a reduced temperature, it may be possible to obtain scintillation breakdowns at much lower current densities.

Conclusions and Recommendations

In conclusion it may be stated that the work reported here indicates that breakdown of surfaces can occur with low-energy particle charging. The scintillations first reported by Malter deserve consideration as a potential source of noise and contamination on synchronous-orbit satellites because they occur on surfaces normally considered to be conductive. The resulting breakdowns have been shown to damage surfaces as the result of their occurrence. These same discharges generate effluents which subsequently can redeposit somewhere else on the satellite. It also has been shown that Malter scintillations cause the motion of 30×10^{-9} C of electrons to distances of several inches above the surface of the vehicle. Aircraft experience indicates that such charge transfers generate signals capable of affecting electronic systems on the vehicle.

Since the present program was very cursory in nature, it is recommended that a more thorough laboratory investigation be made of this discharge process using other materials and more carefully controlled temperatures and illumination. Satellite experiments should be devised to substantiate the laboratory work and to look for unexpected phenomena.

The importance of satellite tests cannot be stressed too strongly. Without them, laboratory work cannot proceed beyond a certain point because the experimenter ultimately must choose between several alternate mechanisms that are possible, depending upon details of the environment that exists in space. Unless such orbital data are generated periodically, the laboratory and analytical work becomes progressively less relevant.

References

[1] Nanevicz, J. E. and Hilbers, G. R., "Titan Vehicle Electrostatic Environment," Stanford Research Institute, Menlo Park, Calif., TR AFAL-TR-73-170, Contract No. F33615-70-C-1406, Project 8428, July 1973.

[2] Nanevicz, J. E., Adamo, R. C., and Scharfman, W. E., "Satellite-Lifetime Monitoring," Stanford Research Institute, Menlo Park, Calif., Final Rept. Contract No. F04701-71-C-0130, P. O. 126192, Project 2611, March 1974.

[3] DeForest, S. E., "Spacecraft Charging at Synchronous Orbit,"

Journal of Geophysical Research, Vol. 77, No. 4, Feb. 1972, pp. 651-659.

[4] DeForest, S. E., "Electrostatic Potentials Developed by ATS-5," AFAL-TR-72-325, 1972 Lightning and Static Electricity Conference Papers, Air Force Systems Command, Wright-Patterson Air Force Base, Ohio, 1972, pp. 150-165.

[5] Feuerbacher, B. and Fitton, B., "Experimental Investigation of Photoemission from Satellite Surface Materials," Journal of Applied Physics, Vol. 43, No. 4, April 1972, pp. 1563-1572.

[6] Knott, K., "The Equilibrium Potential of a Magnetospheric Satellite in an Eclipse Situation," Planetary Space Science, Vol. 20, Pergamom Press, Ltd., London, Aug. 1972, pp. 1137-1146.

[7] Adamo, R. C. and Nanevicz, J. E., "Spacecraft Charging Studies of Voltage Breakdown Processes on Spacecraft Thermal Control Mirrors," published elsewhere in this volume.

[8] Shaw, R. R., Nanevicz, J. E., and Adamo, R. C., "Observations of Electrical Discharges Caused by Differential Satellite Charging," published elsewhere in this volume.

[9] Meulenberg, A. Jr., "Evidence for a New Discharge Mechanism for Dielectrics in a Plasma," published elsewhere in this volume.

[10] Gross, B., "Irradiation Effects in Plexiglas," Journal of Polymer Science, Vol. XXVII, Jan. 1958, pp. 135-143.

[11] Furuta, J., Hiraoka, E., and Okamoto, S., "Discharge Figures in Dielectrics by Electron Irradiation," Journal of Applied Physics, Vol. 37, March 15, 1966, pp. 1873-1878.

[12] Malter, L., "Anomalous Secondary Electron Emission, a New Phenomenon," The Physical Review, Vol. 49, March 15, 1936, p. 478.

[13] Malter, L., "Thin Film Field Emission," The Physical Review, Vol. 50, July 1, 1936, pp. 48-58.

[14] Koller, L. R. and Johnson, R. P., "Visual Observation of the Malter Effect," The Physical Review, Vol. 52, Sept. 1, 1937, pp. 519-523.

[15] McKay, K. G., "Secondary Electron Emission," Advances in Electronics, Vol. 4, Academic Press, Inc., N.Y., 1948, pp. 117-120.

[16] Bruining, H., Physics and Applications of Secondary Electron Emission, Pergamom Press, Ltd., London, 1954, pp. 59-63.

[17] Katz, L. and Penfold, A. S., "Range-Energy Relations for Electrons and the Determination of Beta-Ray End-Point Energies by Absorption," The Reviews of Modern Physics, Vol. 24, Jan. 1952, pp. 28-44.

[18] Tanner, R. L. and Nanevicz, J. E., "Radio Noise Generated on Aircraft Surfaces," Stanford Research Institute, Menlo Park, Calif., Final Rept. Contract AF 33(616)-2761, Project 1267, Sept. 1956.

[19] Nanevicz, J. E., "A Study of Precipitation-Static Noise Generation in Aircraft Canopy Antennas," Stanford Research Institute, Menlo Park, Calif., TR 62, Contract AF 19(604)-1296, Project 1197, Dec. 1957.

SPACECRAFT-CHARGING INVESTIGATION FOR THE CTS PROJECT

N. John Stevens* and Robert R. Lovell+
NASA Lewis Research Center, Cleveland, Ohio

and

Victor Gore[‡]
Communications Research Center, Ottawa, Canada

Abstract

Satellites in synchronous orbit have experienced electronic anomalies that are believed to be caused by discharges from spacecraft surfaces that were differentially charged by the space environment. Because of concern for these environmentally induced anomalies, a program has been undertaken at the NASA Lewis Research Center to assess the impact of the discharge pulses on the joint Canadian-American Communications Technology Satellite (CTS). Whereas this program includes both analytical and experimental investigations, only the experimental results to date are reported in this paper. All of the insulator surfaces tested experiences visible discharged when subjected to an electron beam with energy greater than 10 keV. The discharge rate was found to be a strong function of the current flux. Large areas of the insulator appeared to discharge in each event. The deployable solar-array sample experienced discharges when bombarded on either the cell side or the Kapton side. However, there was no measurable cell performance degradation due to the discharges. The only system tested was the command receiver, which was found to be protected by the band-pass filter.

Presented as Paper SA43 at the 1975 Spring Annual Meeting of the AGU, Washington, D.C., June 17, 1975.
*Aerospace Engineer.
+Chief, Systems Engineering Branch.
[‡]CTS Spacecraft Charging Experimenter.

Introduction

The joint Canadian-American Communications Technology Satellite (CTS) is shown in Fig. 1. It uses a 12-GHz 200-W rf transmitting system and is the first of a new generation of high-power, high-frequency communications satellites. The satellite and its mission have been described in the literature.[1] The launch of this satellite into a synchronous orbit at about 116° W long is scheduled for January 1976.

Satellites in synchronous orbits have been experiencing anomalous electronic switching events, especially when they are in the local midnight-to-dawn portion of their orbits.[2] The data from an experiment on the ATS-5 and 6 satellites have shown that clouds of kilovolt electrons can occur at synchronous altitudes in this quadrant.[3-5] It is believed that these clouds can charge the satellite grounds to potentials that range from a few hundred volts to several kilovolts negative. The range to which the spacecraft grounds can be charged is determined by the areas of the satellite metal surfaces that are in the sunlight. The photoemitted electron current from these sunlit surfaces can balance the incoming electron flux partially and maintain the spacecraft potential within a few hundred volts relative to the space potential.[4] If the spacecraft ground can be charged in this manner, then it must be assumed that the insulators also can be charged and, furthermore, that

Fig. 1 Communications technology satellite (CTS).

the insulator surfaces that are shaded can be charged to the
kilovolt level, even when the spacecraft grounds are maintained
at the few hundred volt level. The satellite can be differentially charged in this manner. When the satellite surfaces are
charged to the kilovolt range, a discharge can occur, and the
resulting pulse of electromagnetic energy can trigger sensitive
electronic logic systems. This phenomenon of the environment
charging the satellite surfaces has been given the name
spacecraft-charging.

The CTS was designed in the 1970-71 time period when the
spacecraft-charging phenomenon barely was recognized by project
personnel. The satellite incorporated design features which
normally were used at that time for lightweight satellites.
Thermal blankets were used to close the top and bottom spacecraft body openings and, in addition, solar cells, optical
solar reflectors, and silvered Teflon were used on the satellite exterior.

The unique feature of the satellite is the size and construction of the main solar-array wings, each 1.2x0.76 m long,
which have the solar cells mounted on a 3-mil-Kapton-fiber
glass composite substrate. The satellite will be three-axis
stabilized, and these solar arrays will track the sun. Therefore, this satellite will have large areas of insulator surfaces that will be shaded when the satellite is in the predawn
quadrant, and these surfaces can be charged by the plasma
clouds generated by the geomagnetic substorms.

The concern for possible spacecraft-charging anomalies on
the CTS led to a program at the NASA Lewis Research Center to
assess the impact of environmentally induced discharges on the
satellite electronic systems. The objective of this program is
to generate design requirements and test specifications for
subsystem and spacecraft level tests. The approach used is to
conduct experimental investigations to determine the material
response to the simulated substorm environment, to conduct analytical investigations to establish the test specifications and
design criteria, and to conduct spacecraft systems tests to determine the systems response to the discharge pulses from the
satellite surfaces. In this paper, only the experimental investigations of the material response and the results of the
command receiver system test will be discussed.

Experimental Investigation

Facility

The facility used to determine the satellite material response is shown schematically in Fig. 2. All testing is per-

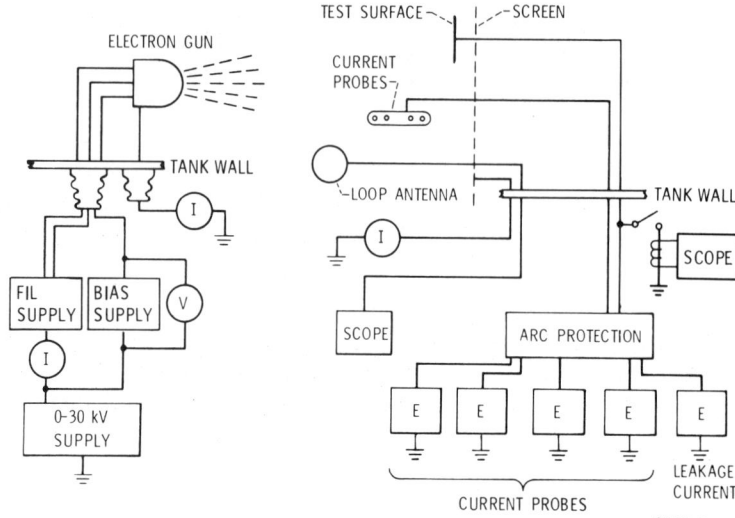

Fig. 2 Spacecraft-charging experimental facility.

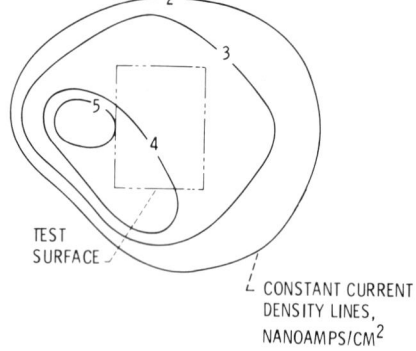

Fig. 3 Typical uniformity at test plane.

formed in a 1.8-m-diam x 1.8-m-long (6-ft-diam x 6-ft long) vacuum chamber capable of operating at a pressure in the range of 1–5×10^{-7} torr. The grounded metallic wall forms the boundary condition for all tests. A divergent electron beam is generated from a hot-wire filament by means of a curved accelerating screen. The filament is biased to the desired voltage level, and the accelerating screen is kept at ground potential. The electron generator is surrounded by a ground screen to minimize the stray electric fields from the high-voltage system.

The beam from this source is fairly uniform. Typical results obtained from grounded metallic current sensors, located at the test plane 1 m from the source are shown in Fig. 3. The typical sample size and location also are shown in Fig. 3. The

Table I Materials surveyed

Silvered Teflon 5 mils
  Silver layer isolated
  Silver layer grounded
Solar array segment
  Deployable array segment (27 cells)
  (Front and back irradiated)
Optical solar reflectors
  Silver layer isolated
  Silver layer grounded
Thermal blankets
  Metal layers grounded

test surface, as shown in Fig. 2, is located on the electron source centerline. The current probes surround the specimen so that continual beam-current monitoring is possible during the test. The loop antenna picks up the discharge activity and displays the voltage profile from the discharge on a fast storage oscilloscope. The screen behind the sample is luminescent and can be viewed to watch the surface-charging activity as well as to check beam uniformity. The back of the test sample is connected to ground through a Keithley digital electrometer (model 610) to measure the leakage current through the sample.

This current monitoring line also can be shorted directly to ground so that a current probe connected to the oscilloscope can be used to obtain the transient current pulses from discharges.

Test Procedure

The response of the material to an electron flux was to be determined for the types of surfaces being used on the satellite. These materials are shown in Table I. The majority of the investigations were conducted on 15 x 20-cm samples of 0.013-cm silvered Teflon and the 9.6 x 20.8-cm sample of the deployable solar array. Brief survey testing only has been conducted on the optical solar reflectors and the thermal blankets.

The usual test was to subject a sample to bombardment from electrons accelerated through 5, 10, 15, and 20 kv and at electron beam current densities of 1 and 10 nA/cm^2. The sample was subjected to this impingement for 16 min at each voltage and current level. The leakage current, the transient current

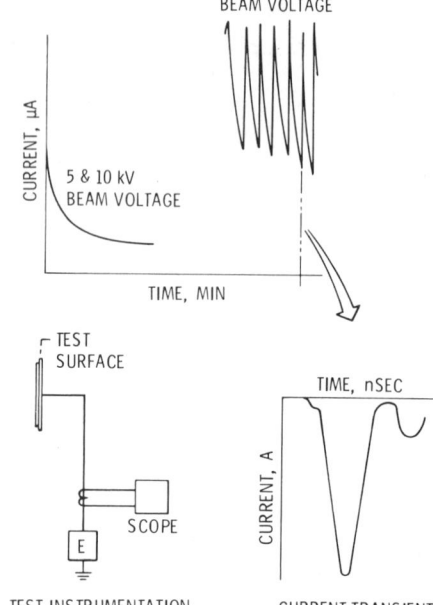

Fig. 4 Typical sample current behavior.

pulses, the antenna pickup, and the number of discharges were recorded at each step.

Test Results

Metallized Insulators

The leakage electron current as a function of time under electron bombardment is shown pictorially in Fig. 4. When the beam voltage is less than 10 kv, there are no visible discharges on the insulator surface, and the electron current through the material is shown in the left-hand portion of the figure. This curve has the characteristics of a capacitor charging circuit with a nonlinear resistance in series with the capacitor. The steady-state value is due to the leakage through the insulator when the front face is fully charged and can be used to estimate the effective bulk resistance. For the silvered Teflon samples, it requires about 10-20 min to reach this value, and the steady-state leakage current represents a bulk resistivity of about 10^{16} ohm-cm.

When the beam voltages are greater than 10 kv and at either current density, there are visible discharges over large

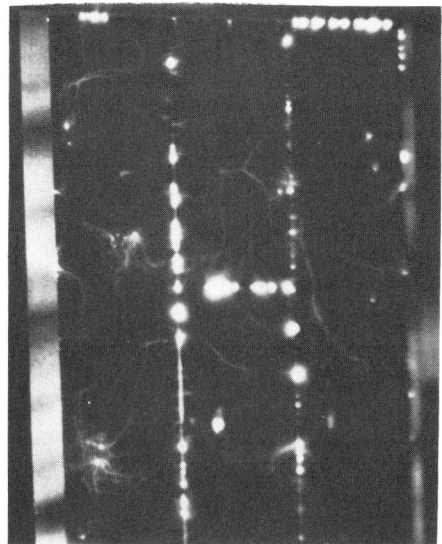

Fig. 5 Silvered teflon tape test specimen (electrically floating silver layer).

portions of the surface. The electron leakage current to ground is modified as shown in the right-hand portion of Fig. 4. When the electron beam first is applied, the leakage (or displacement) current jumps to a high value and starts to decay as before. Within 4 min, there will be a discharge that causes a large transient pulse. The leakage current then returns to a high value and starts to decay again. The oscilloscope trace of the transient is shown in the lower right-hand side of Fig. 4. The transient pulse has been measured as high as 100 amp and lasts between 200 and 300 nsec. It represents an electron current flow from ground to the substrate, which is in opposition to the electron leakage current.

The luminescent screen was used extensively during the testing. The shadow pattern on the screen was observed to grow as the test surface was charged. When a discharge occurred, the shadow collapsed to its original size and started growing again. This behavior supports the belief that a large surface area is discharged in a single pulse. The discharge activity, as determined by the discharge counter, continues throughout the test at a rate dependent primarily upon the incident electron flux. For the silvered Teflon samples the discharge rate is about 4/min at beam currents of 10 nA/cm^2 and at 1 nA/cm^2, the rate is about 1/5 min.

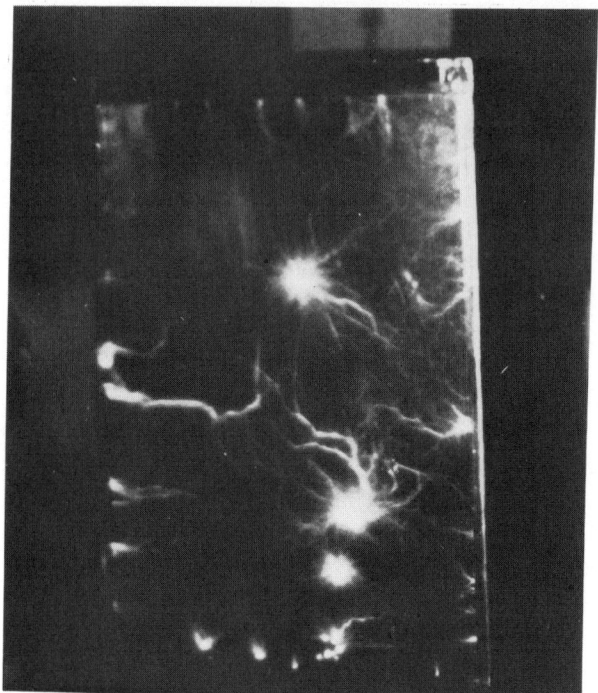

Fig. 6 Silvered Teflon test specimen (ground silver layer).

The visible discharges viewed on the sample of silvered Teflon are shown in Fig. 5. This picture is a time exposure and has captured two major discharges. The sample is three strips of 0.013-cm-silvered Teflon (Teflon side under bombardment), 5 cm wide x 20 cm long, mounted on the aluminum substrate with Kapton tape having adhesive on both sides. This mounting technique is identical with the actual satellite mounting. Whereas this technique insures that the silver will stay on the Teflon through the anticipated severe thermal cycles, it does electrically isolate the metal layer. As a result of this isolation, there are spot discharges at all tape edges, and each spot represents silver being vaporized. Substantial amounts of silver can be lost in this manner. For example, 12 percent of the silver was lost in 43 hr of testing at a beam voltage of 20 kv and beam current density of 10 nA/cm^2.

Grounding of the silver layer, however, does not solve all of the problems, as is shown in Fig. 6. This is a silvered Teflon sample with conductive adhesive holding the silver layer to the substrate. In this case, there are punch-through dis-

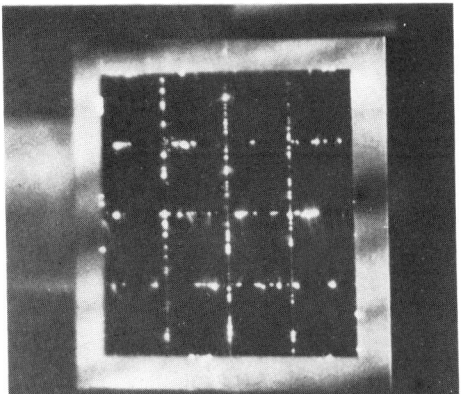

Fig. 7 Optical solar reflector test specimen (electrically floating silver layer).

charges (dielectric to metal) with accompanying surface streamers. The streamers, which also were observed in the floating-silver-layer samples (Fig. 5), are believed to be due to the discharge of large areas of the Teflon surface. All that is accomplished by the grounding of the silver is a reduction in the silver loss. Once the insulator is punctured, the discharge rate approaches that of the ungrounded metal samples.

The phenomenon of material loss from an isolated metal layer has been observed in the optical solar reflector (OSR) testing as well. When the OSR's were mounted to the substrate with double-sided Kapton tape, there were spot discharges and loss of the metal layer (Fig. 7). When the OSR's were mounted with a conductive adhesive, there were only glow discharges from the quartz and no metal loss. The thermal blanket test samples had grounded metal layers, and although discharges were visible from the surface, there was no material loss. It appears that the grounding of the thermal blanket metal layers does not prevent discharges but does stop the metal loss associated with the discharge.

Solar-Array Segment

The flexible solar-array segment also has been tested with both the cell side and the Kapton side under bombardment. Figure 8 shows the visible discharges with the Kapton side under bombardment. Note that the majority of the discharges occurred at the cell interconnects, which were at ground poten-

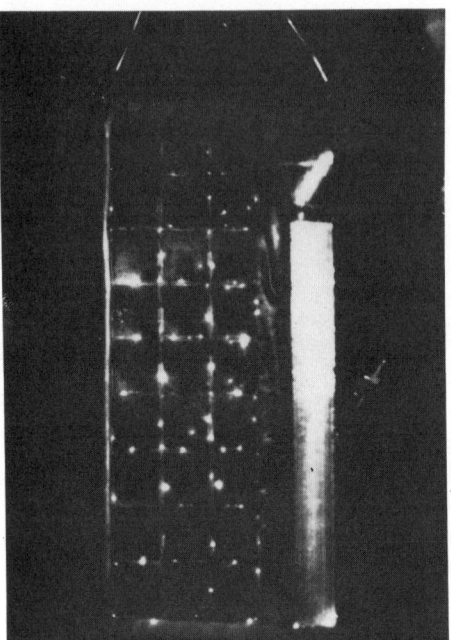

Fig. 8 Deployable solar-array test specimen (electron bombardment on Kapton side).

tial during this test. The visible discharges occurred at beam voltages of 15 kv and at both current densities. The discharge rate again depended primarily on the current density. This sample was tested with the Kapton side bombarded for 6 hr without significant change in the discharge rate. The voltage-current characteristics of the solar cells were measured before and after this test; no degradation of the cell performance was observed. There was also no material loss observed at the interconnects due to the discharges. The bombardment of the cell side of the solar-array segment produced glow-type discharges from the cover glass when the beam voltage was at or above 15 kv and at both current densities. There was neither degradation of cell performance nor material loss from this test.

Command Receiver Test

One of the samples that was tested in the experimental investigation was a segment of a CTS communications antenna called the belt antenna. This antenna is essentially a large

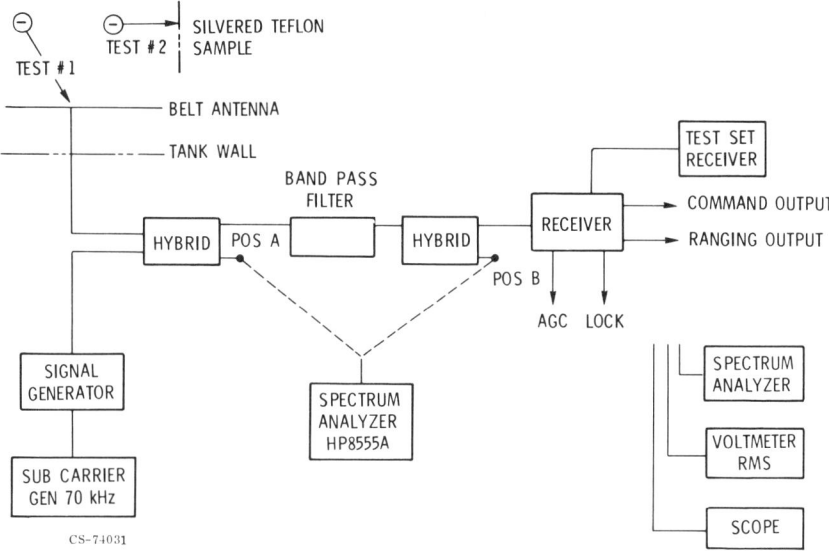

Fig. 9 Belt antenna: command receiver test.

printed circuit board with both telemetry and command metal patches deposited on Teflon-filled fiber glass. The sample of the antenna discharged during survey tests, and there was concern that these discharge pulses could disrupt the command receiver. Therefore, two tests were run on the engineering model of the command receiver system as shown in Fig. 9. The belt antenna was placed inside the vacuum chamber and bombarded directly in the first test. The receiver system consisting of the two hybrids; a band-pass filter, and the receiver were outside the tank. A signal was provided to lock up the receiver. The pulses from the discharges on the antenna were conducted through the lines and observed at the output of the first hybrid. The band-pass filter, however, absorbed the pulses, so that there was no effect on the receiver output. The band-pass filter was included in this system for reasons other than spacecraft-charging, but it also does protect the receiver from discharges on the antenna. A second test was run with the antenna as the receiver for discharges from a silvered Teflon sample. The results of this test were the same as the first.

Summary of Results

All of the insulator samples tested in the experimental portion of this program discharged when subjected to electron bombardment. There were visible discharges on the insulator surfaces when the beam voltage was above 10 kv. The discharge rate depended primarily on the incident current flux. The observations of the visible discharges, the changes in the shadow pattern on the luminescent screen behind the test sample, and the characteristics of the leakage current all indicated that large areas of the sample were discharging in a single event. The transient current pulse from a discharge was many amperes in magnitude and represented an electron-current flow in the direction opposite to the leakage current.

For the silvered Teflon samples, isolating the metal layers results in spot discharges at all edges and loss of metal. Grounding of the metal layer results in reduced silver loss, but little reduction in discharge activity, once the insulator layer has been punctured. An ungrounded metal layer in the optical solar reflectors also causes the metal-loss type of discharges. The grounding of the metal layers in the thermal blankets prevents the metal loss but does not prevent discharges.

Discharges were observed when the solar-cell segments were tested. Glow discharges were observed when the cell sides were bombarded with electrons having greater than 10-kev energies. Spot discharges, primarily at the grounded interconnects, were observed when the Kapton (or rear) side of the array segment was bombarded with electrons of greater than 10 kev. There was no cell deterioration measured as a result of these tests.

The only CTS system tested was the command receiver. This system was protected by the band-pass filter, which is an integral part of the flight system. There is still concern for possible discharges on the CTS. This concern has resulted in the inclusion of a transient-event counter on the satellite to monitor pulses at three locations in the main harness.

References

[1] Franklin, C. A. and Davidson, E. H., "A High-Power Communications Technology Satellite for the 12 and 14 GH_z Bands," AIAA Paper 72-580; also, AIAA Progress in Astronautics and Aeronautics: Communications Satellite Systems, Vol. 32, edited by P. L. Bargellini, The MIT Press, Cambridge, Mass., 1974, pp. 87-121.

[2] McPherson, D. A., Cauffman, D. P., and Schober, W., "Spacecraft Charging at High Altitudes - The SCATHA Satellite Program," AIAA Paper 75-92, 1975, Pasadena, Calif.

[3] DeForest, S. E. and McIlwain, C. E., "Plasma Clouds in the Magnetosphere," Journal of Geophysical Research, Vol. 76, June 1974, pp. 3587-3611.

[4] De Forest, S. E., "Spacecraft Charging at Synchronous Orbits," Journal of Geophysical Research, Vol. 77, Feb. 1972, pp. 651-659.

[5] Bartlett, R. O., DeForest, S. E., and Goldstein, R., "Spacecraft-Charging Control Demonstration at Geosynchronous Altitude," AIAA Paper 75-359, 1975, New Orleans, La.

SPACECRAFT-CHARGING ANALYSIS AND TEST FOR ENVIRONMENTALLY INDUCED EMI

A. Krausz*

TRW Systems Inc., Redondo Beach, Calif.

Abstract

Spacecraft that are exposed to the charged particle plasma resulting from intense geomagnetic substorm activity will assume a sufficiently high electrostatic potential to cause spurious arc discharge at locations of maximum stress. An analysis of the resulting electromagnetic interference (EMI) is presented herein and is based on arc characteristics obtained from a laboratory simulation of arc discharges from thermal insulation blankets. The conclusion is reached that spurious discharges do not present a threat to the spacecraft's onboard electronic equipment unless the discharge arc strikes a wire or a wire shield which can couple the discharge energy to a low level circuit.

Introduction

The charged particle environment during intense geomagnetic substorm activity and the resulting electrostatic charge accumulated by a spacecraft have been studied by De Forest,[1] Rosen,[2] and Inouye.[3] It has been shown that potential difference between the exposed surface of the spacecraft and the structure which serves as the electrical "ground" plane may

---

Presented as Paper SA45 at the American Geophysical Union National Meeting, Washington, D. C., June 1975.
   *Manager, Systems Engineering Staff, Defense and Space Systems Group.

reach values of 10-20 kV. This potential difference is great enough to cause intermittent arcing and an attendant sudden change in potential difference to cause electromagnetic interference (EMI) that upsets the vehicle's electronic circuits or causes permanent damage due to the energy contained in the arc. This paper describes some recent tests to determine the amplitude and spectral characteristics of the arc discharge and discusses the interference voltages that can be expected at the input to typical electronic equipments.

Charge Buildup

Exterior surfaces of spacecraft consist of various materials such as solar-cell panels, second-surface mirrors, or aluminized Mylar blankets to control thermal radiation properties. The electrostatic potential of such surfaces when exposed to the space plasma results from the process illustrated by Fig. 1.

Charged particles from the space plasma are absorbed by the spacecraft until its potential with respect to the plasma is such that the net current between the spacecraft and the plasma is zero. If there are no lateral surface currents to equalize the charge among different surfaces, exterior surfaces that are illuminated will reach an electrostatic potential of perhaps 100 V negative, since photoelectrons that leave the surface will balance the predominant electron current, which reaches the surface from the environment. Dark surfaces, on the other hand, will charge to a large negative voltage of several kilovolts until incoming electrons are repelled by the potential barrier due to the electrostatic charge. A possible discharge location is illustrated in Fig. 2, which shows a thermal blanket covering an area of exposed structure. Electrons, which are deposited on the exterior layer of the blanket, will give rise to a voltage of perhaps 10 kV or higher between it and the structure. If the aluminized layers of the thermal blanket are isolated electrically, a significant potential difference also exists between adjacent layers and between any layer and the structure.

Discharge Characteristics

The potential difference between a charged exterior dielectric surface and structure or between a metalized layer and structure during severe substorms is high enough to cause

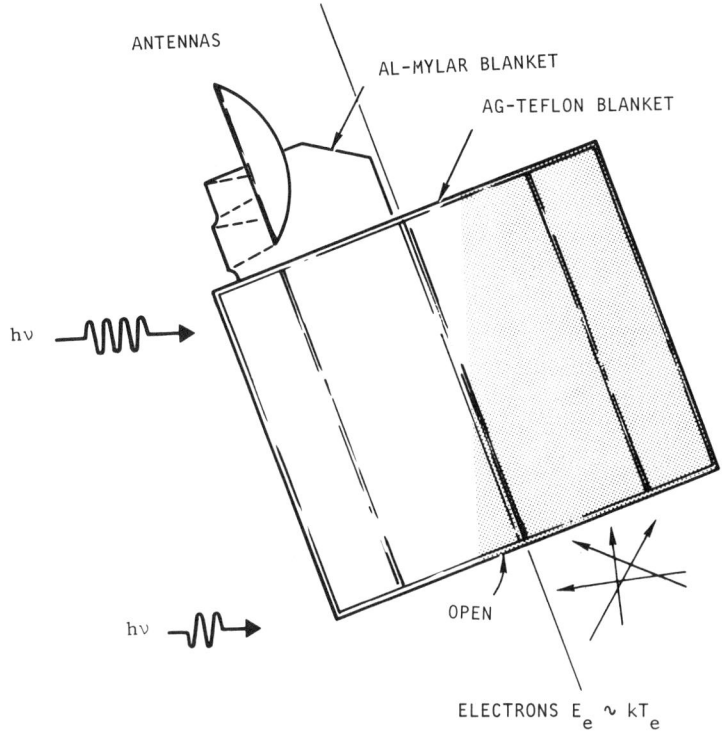

Fig. 1 Satellite-charging.

ILLUMINATED SIDE
PHOTOELECTRON CURRENT NEARLY NEUTRALIZES CHARGE, SURFACE POTENTIAL ≃ NEG 100-300V

DARK SIDE
NO PHOTOELECTRONS, SURFACE CHARGES TO V∼-kT$_{e/e}$ OR NEG SEVERAL kV.

INTERIOR
NO PHOTOELECTRONS, CAN CHARGE TO ∼-kT$_{e/e}$ OR SEVERAL kV DUE TO ELECTRONS ENTERING OPENINGS.

arc discharge at some point between the charged surfaces. Thus, both dielectric-to-metal and metal-to-metal arc discharges can be expected. The amplitude and waveform of the arc current are functions of the stored energy, the location of the arc, the geometry, and the conductivities that are involved. In order to gain some understanding of the electrical characterisitics of such arcs, representative samples of thermal blanket materials were exposed to a stream of 10 keV electrons in a vacuum chamber.[4] Thermal blankets were chosen for this test over other materials, such as solar panels or second-surface mirrors because, in general, the electrostatic

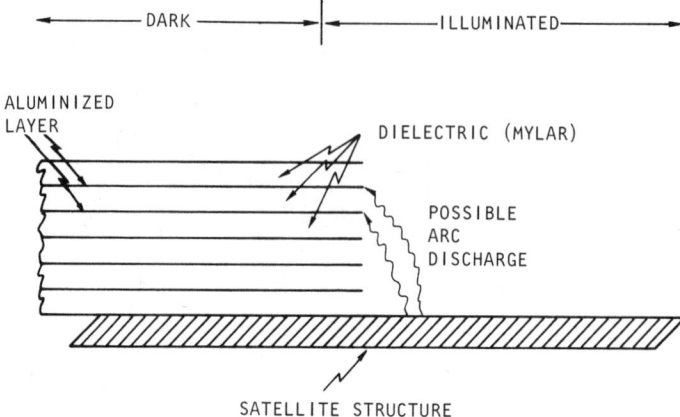

Fig. 2 Thermal blanket discharge.

energy that is released by each arc is greater for differentially charged thermal blankets than for other surface elements. The resulting dielectric breakdown currents were determined by means of small inductive and capacitance probes as shown in Fig. 3.[4] For metal-to-metal surface breakdown from a 1000-cm^2 thermal blanket, the arc currents had peak amplitudes of 200-1000 A with rise times of less than 100 nsec and typical pulse duration of 150-500 nsec. Dielectric-to-metal arcs also were observed but had lower amplitude by at least one order of magnitude and much greater pulse duration. Therefore, since they would be much less effective in generating EMI, they will not be considered further.

EMI Susceptibility

The susceptiblity of electronic equipment to EMI depends on the detailed circuit design, the input impedance, and the physical configuration of the equipment. EMI susceptibility limits of electronic equipment are controlled by an electromagnetic compatibility (EMC) plan or applicable government specifications such as MIL-STD-461 and MIL-STD-1541. Although requirements must be tailored to specific applications, the following are typical:

1) No malfunction when exposed to an E field of 5 V/m from 14 kHz to 10 GHz.

2) No false response of digital circuits when exposed to an arc discharge of 10 kV at 30 cm from a Tesla coil.

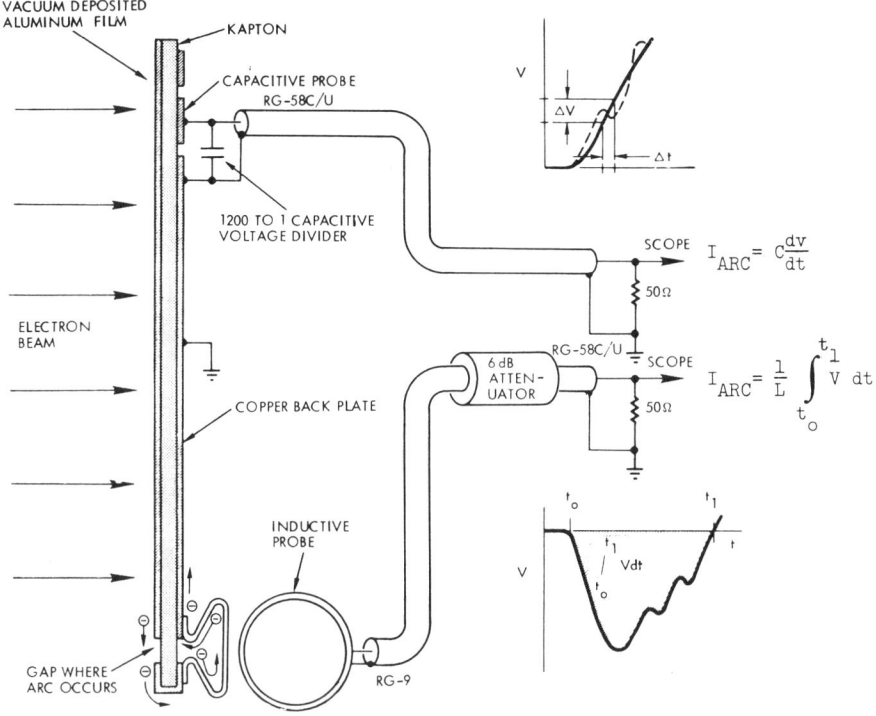

Fig. 3   Observation of metal-to-metal arc.

3) Equipment shall not be susceptible to noise voltages of less than 1 V pp at frequencies up to 400 MHz appearing on primary power lines or to 50 mV rms on secondary power lines.

4) Digital circuits shall not respond to pulses of less than 1 V and rise time less than 10 nsec.

5) Receivers shall reject out-of-band signals such that the resulting out-of-band receiver output voltage is at least 60 dB lower than the desired in-band output.

Grounding, shielding, filtering, and gating techniques are used to meet these requirements.

EMI Analysis

If the location, amplitude, and waveform of the arc current are known, its interference effect on electronic circuits may be determined. Accurate prediction of the interference voltage or current induced in any receptor circuit would require solution of an extremely complicated boundary value problem and an exact knowledge of the interference source. In practice, however, it is sufficient to obtain an upper limit of the induced voltage or current at any critical circuit location as a function of frequency (i.e., the spectral intensity) that can be expected due to arcing. The problem, therefore, may be simplified by considering separately the following three methods of interference propagation: 1) radiation of electric and magnetic fields from the arc, 2) the effect of ground replacement currents, and 3) injection of arc currents into a cable harness.

The analysis is simplified greatly by considering only amplitude vs frequency characteristics and ignoring all phase relationships. Experience has shown that this gives results which may be larger than actual measured voltages or currents by no more than 10 dB. This degree of possible error is acceptable for design purposes.

For the case of environmentally induced EMI, only metal-to-metal arcs from 1000-cm^2 thermal blankets with floating metalization layers are considered herein, since this represents a reasonable worst-case situation. As mentioned previously, although no reliable laboratory measurements of dielectric-to-metal arc currents are available, the amplitude is assumed to be sufficiently low that such arcs do not represent a threat to a spacecraft that meets the usual electromagnetic susceptibility requirements. The results of our effort to predict interference levels due to the radiation field, ground replacement currents, and cable-injected currents from discharge of a 1000-cm^2 thermal blanket to structure now will be described.

Radiated Interference

The radiated E and H field caused by the arc discharge is attenuated by the spacecraft structure and the electronic equipment enclosures. Figure 4 shows how much attenuation is provided by a solid .013-cm shield. In most spacecraft, more than this amount of shielding is provided automatically by the structure and the equipment enclosures. Unless the arc is visible through an aperture clearly, the field strength of the arc source therefore must be greater than 10^6V/m to produce a field in excess of the 1-V/m susceptibility threshold inside the spacecraft. Although no direct field measurements were possible in the vacuum chamber, the estimated field strength

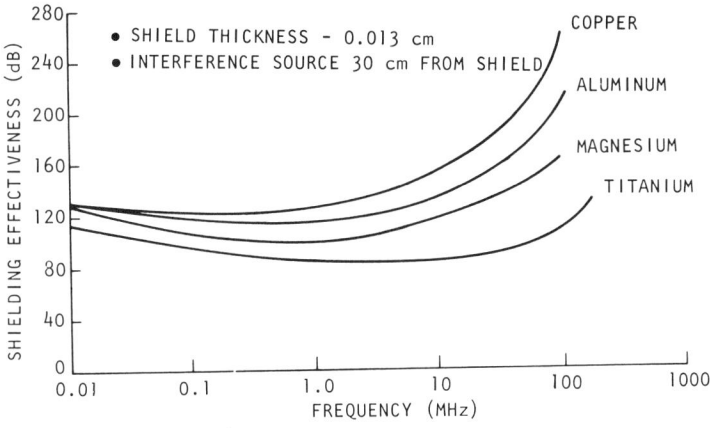

Fig. 4 Shield effectiveness.

of the thermal blanket discharge is several orders of magnitude lower than this.

Effect of Ground Replacement Current

In conjunction with direct arc conduction currents involving stored charges, there are displacement currents between the arc and the spacecraft which result from capacitive and mutual inductance effects. The magnitude of such displacement or redistribution currents is strongly dependent on the rate of change of the driving source current.

Figure 5 illustrates a test that was conducted on the full-scale structural model of a spacecraft to measure the internal fields due to replacement currents. The test also can

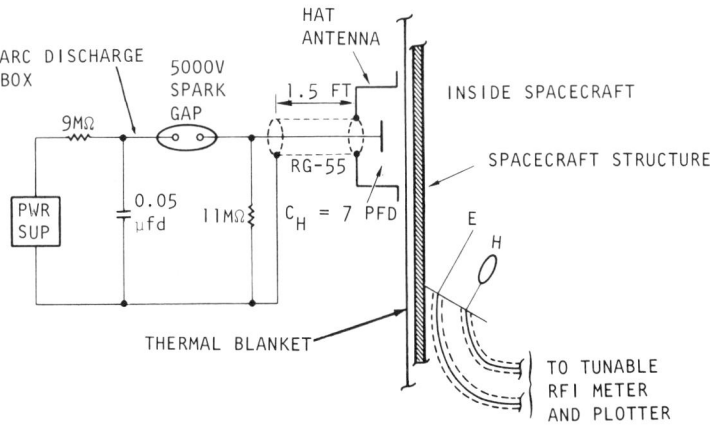

Fig. 5 Test configuration: arc discharge field.

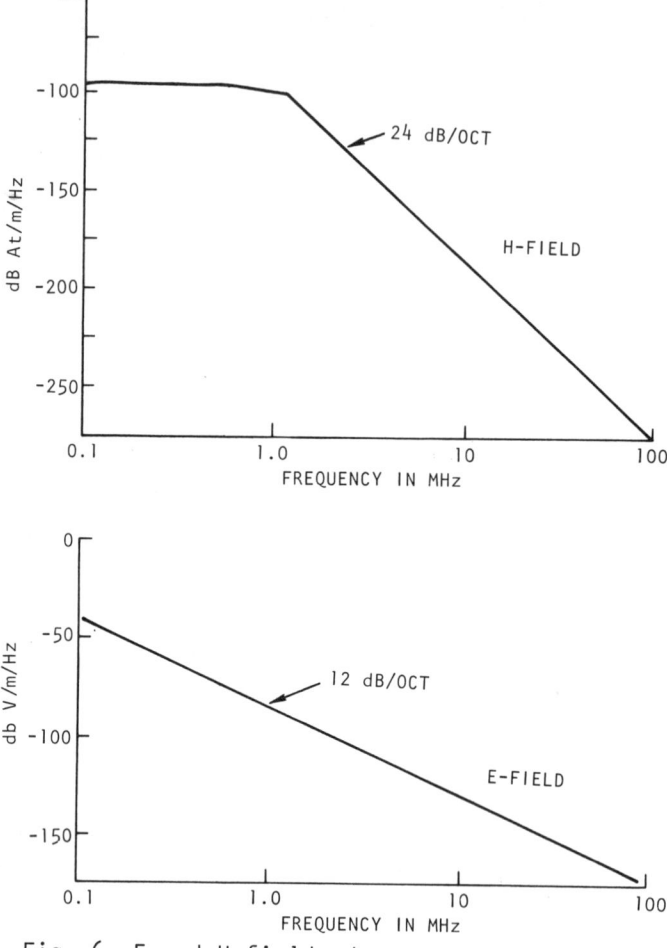

Fig. 6 E and H fields due to arc discharge.

be interpreted as a confirmation of shielding effectiveness when arc current flows to the shield. The current that flows in the ground plane when a metal-to-metal arc occurs is simulated by means of a hat antenna connected to a high-voltage power supply through a spark gap. The charging current of C_H simulates the discharge of a small thermal blanket. The resulting field measurements are shown in Fig. 6 and have been corrected for the difference in amplitude of the antenna current and the blanket discharge current which, based on the measurements of Ref. 4, has been represented as a double exponential pulse that reaches its peak amplitude of 1000 amps in 200 nsec. Although the bandwidths of the EMI meters are not indicated, the internal fields are below threshold sensitivities of electronic equipment that meets MIL-STD-461.

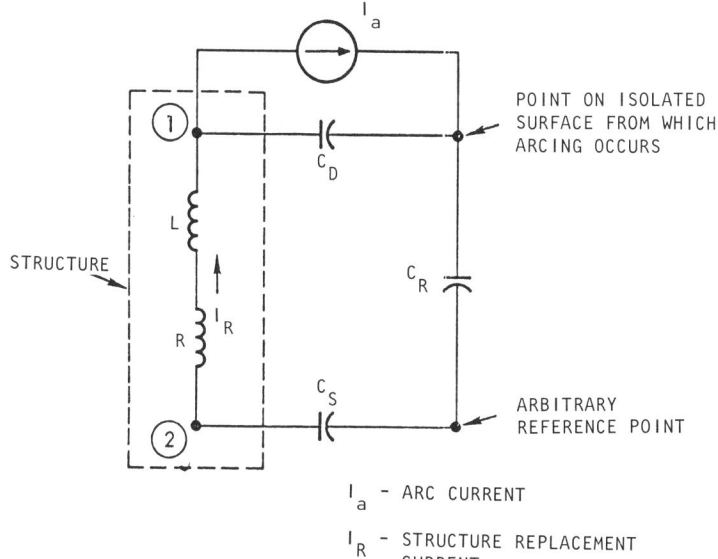

Fig. 7 Equivalent circuit for calculation of replacement current.

Mathematical modeling also has been used to estimate the magnitudes of replacement currents at various locations. A highly simplified equivalent circuit for modeling the effect of arc discharges from a surface with capacity C_D to structure at point 1 is shown in Fig. 7. The equivalent inductance, L, for various structure elements of the spacecraft and the capacitance to free space, was obtained by using textbook equations for circuit equivalence based on isolated cylinders and short transmission lines. Calculations for an actual spacecraft give peak replacement currents of less than a few milliamperes when a 1000-cm^2 surface charged to 10 kV is discharged, assuming any discharge duration of from 1-1000 nsec. Such currents are insufficient to produce interference in the usual electronic units but may be picked up by special probes.

Direct Arcing to a Cable

When the environmentally induced electrostatic field, which terminates on a cable or shield, also creates an area of maximum stress at the cable, the arc will strike the cable or its shield and cause a large localized current. This current will generate a voltage transient on the line that is struck and cause current transients in nearby wires and shields, which are inductively coupled. The magnitude of these transients depends on the geometry of the cable and the terminating impedances, as well as the characteristics of the arc source and, therefore, is difficult to determine. Such voltage or current transients, however, are sufficiently large to cause upsets in digital circuits and low-level sensors. This has been demonstrated amply in the laboratory by striking an arc from a Tesla coil to certain wire bundles that carry low-level signals. Since this appears to be the only manner in which upsets can be produced; all wire bundles, even if shielded, should be kept away from large surfaces that may become charged with respect to the ground plane during substorm activity. If this cannot be done, multiple shields, which are grounded at frequent intervals to reduce the inductive impedance, should be used.

Conclusion

Spacecraft electronic equipment, which meets the usual EMI susceptibility requirements, is not affected adversely by environmentally induced arc discharges from small (less than 1000-cm^2) external surfaces unless the arc strikes a wire bundle which serves low-level circuits. Since environmentally induced EMI can be due only to spurious discharge of accumulated electrostatic charge, the obvious method for preventing EMI is to ground isolated elements of the spacecraft. This means that all conductive surfaces that are exposed to the space plasma must be electrically connected together. Where this cannot be done, a leakage path for electrostatic charges must be provided in order to prevent metal-to-metal arc discharge. In addition, all low-level circuits should be designed to have bandwidth no greater than required for their function. A practical method for verifying, prior to flight, that all exposed surfaces are adequately grounded and all wires are remote from high-stress locations has yet to be developed.

References

[1] De Forest, S. E., "Spacecraft-Charging at Synchronous Orbit," Journal of Geophysical Research, Vol. 77, Feb. 1, 1972, pp. 651-659.

[2] Rosen, A., "Large Discharges and Arcs on Spacecraft," _Astronautics & Aeronautics_, Vol. 13, June 1975, pp. 36-44.

[3] Inouye, G. T., "Spacecraft-Charging Model," AIAA Paper 75-255, January 20-22, 1975, Pasadena, Calif.

[4] Hoffmaster, D. K. and Sellen, J. M. Jr., TRW Systems Inc., Redondo Beach, Calif. Rept. 4351.3. 74-51, Nov. 18, 1974.

Index to Contributors to Volume 47

Adamo, R. C., *Stanford Research Institute*.............. 61, 225, 247
Balmain, K. G., *University of Toronto* 213
Bunn, M. H., *U.S. Air Force Space and Missile Systems Organization*...... 45
Chappell, C. R., *NASA Marshall Space Flight Center* 89
DeForest, S. E., *University of Alabama* 77, 169
Goldstein, R., *Jet Propulsion Laboratory*..................... 169
Gonfalone, A., *European Space Agency, The Netherlands* 159
Gore, Victor, *Communications Research Center* 263
Grard, R., *European Space Agency, The Netherlands*............... 159
Hanser, F. A., *Panametrics, Inc.* 31
Hoffmaster, D. K., *TRW Defense & Space Systems* 185
Hunerwadel, J. L., *Panametrics, Inc.* 31
Inouye, G. T., *TRW Defense & Space Systems* 103
Katz, L., *Air Force Geophysics Laboratory*................. 31, 121
Krausz, A., *TRW Systems, Inc.* ... 277
Kremer, P., *University of Toronto* 213
Lehn, William, *Air Force Materials Laboratory* 3
Lennartsson, Walter, *NASA Marshall Space Flight Center* 89
Lovell, Robert R., *NASA Lewis Research Center*............. 3, 263

McPherson, D. A., *The Aerospace Corporation*................... 15
Meulenberg, A., Jr., *COMSAT Laboratories*................... 237
Morel, P. R., *Panametrics, Inc.*..... 31
Nanevicz, J. E., *Stanford Research Institute*................. 61, 225, 247
Orszag, M., *University of Toronto* 213
Pavel, A. L., *Air Force Geophysics Laboratory*................. 31, 121
Pedersen, A., *European Space Agency, The Netherlands*............... 159
Pike, C. P., *Air Force Geophysics Laboratory* 3, 45
Reasoner, D. L., *NASA Marshall Space Flight Center* 89
Rothwell, P. L., *Air Force Geophysics Laboratory*................. 31, 121
Rubin, A. G., *Air Force Geophysics Laboratory*................... 121
Sanders, N. L., *TRW Defense & Space Systems* 77
Schober, Wayne, *Space and Missile Organization* 3, 15
Sellen, J. M., Jr., *TRW Defense & Space Systems* 185
Sellers, B., *Panametrics, Inc.*...... 31
Shaw, R. R., *Aerojet Electrosystems Company* 61
Stevens, N. John, *NASA Lewis Research Center* 3, 263
Vogl, J. L., *TRW Defense & Space Systems* 77
Whipple, E. C., Jr., *NOAA Environmental Research Laboratories* 135

Subject Index
for Volume 47

Active potential control
.... 154,166,170
Anomalies
 environmental............ 45
 failures 52
Anomaly correlation with
spacecraft model........... 104
Arc discharge
 analysis............. 202,240
 cables 285
 measurements 189,213,225
 237,247,263
 spectrum 218
 theory 205
ATS-5,6
 data 77,89,93,113,153,169
 experiments.............. 92
Auroral ovals 49
Barrier potential 153
Charge deposition 185,240
Charge detectors............ 63
Charging electrical effects
analyzer 24
Circuit upset anomalies....... 17
Communication technology
satellite (CTS)............. 263
Computer
 calculations 110,122,138
 particles................ 123
Conductivity, bulk 108,199
Correlation
 magnetic activity.......... 70
 study................... 50
Current profile 80
Defense Meteorological
Satellite Program (DMSP)
 photographs 51
Defense Satellite Communication
System (DSCS)
 project 45
 II spacecraft 105
Design criteria.............. 12
Dielectric surface potential ... 195
Dielectric-to-metal arcs 213
Directional particles 7
Discharge
 characteristics 278
 command receiver sets..... 272
 detectors 63
 mechanism bilayer model .. 237
 observations 64
 solar array.............. 271
Electric field detector 26
Electric stress levels......... 198
Electrical discharges in orbit ... 61
Electrical network model 103,109
Electromagnetic interference
 analysis................ 281
 radiated................ 282
 susceptibility 280
Electron currents 80,91
 distribution.............. 81

 microscope 214
 temperature 80,81
Electron and ion beam
experiments................ 179
Environmental
 data.................... 77
 definition program 8
 induced anomalies......... 45
Field-aligned fluxes.......... 82
Field emitter 165
Flight experiments 11
Floating potential....... 130,161
Geomagnetic activity 70
Geosynchronous satellites
 77,89,103,169
Ground replacement currents 283
Kapton resistivity 118
Langmuir probe theory...... 126
Langmuir theory 127,130
Local time
 dependence............. 68
 distribution anomalies...... 90
 distribution charging....... 98
Logic upsets 90
Material development........ 10
Material response....... 185,267
Magnetic index 47
Malter discharges 247
Maxwell-Boltzmann distributions
 81,113
Maxwellian distribution 150
Metal-to-metal arcs 185,202
Mirrors, second surface...... 225
Monitors.................. 24
 charging 119,274
Monoenergetic distribution... 151
Omnidirectional measurements 78
Particle and fields instrumenta-
tion 27
Particle gun 27
Plasma density 153,161
Plasma encounter 97
Photoconduction 200
Photoelectron sheath 135
Photoelectron temperature ... 153
Photosheath 128
Pitch angle distribution....... 84
Plasma sheath 121,136
Potential buildup-time
dependence............... 124
Potential distributions in plasma
sheaths 137
Poissons equation 151
Proton current.......... 80,91
Quartz crystal microbalance ... 26
Research and technology
program 3
Resistivity
 bulk................... 199
 dielecfric 268
 effects of UV............ 201
 non-linear bulk 108